FROM X-RAYS
TO QUARKS

Modern Physicists and Their Discoveries

W. H. FREEMAN AND COMPANY *San Francisco*

FROM X-RAYS TO QUARKS

Modern Physicists and Their Discoveries

Emilio Segrè
UNIVERSITY OF CALIFORNIA, BERKELEY

Original edition, *Personaggi e Scoperte nella Fisica Contemporanea,* published by Mondadori, Milan, 1976.

Sponsoring Editor: Arthur C. Bartlett; *Project Editor:* Patricia Brewer; *Copyeditor:* Susan Weisberg; *Designer:* Sharon Smith; *Production Coordinator:* Linda Jupiter; *Illustration Coordinator:* Cheryl Nufer; *Compositor:* Bi-Comp, Inc.; *Printer and Binder:* The Maple-Vail Book Manufacturing Group.

Library of Congress Cataloging in Publication Data

Segrè, Emilio.
  From x-rays to quarks.

  Translation of Personaggi e scoperte nella
fisica contemporanea.
  Bibliography: p.
  Includes indexes.
  1. Physics—History. 2. Physicists—Biography.
I. Title.
QC7.S4413   530'.09   80-466
ISBN 0-7167-1146-X
ISBN 0-7167-1147-8 pbk.

9 8 7 6 5 4

Contents

Preface

This book is based on lectures I gave at the University of California, Berkeley, at the University of Chicago, and at the Accademia Nazionale dei Lincei in Rome. The many flattering requests I received from audiences expressing a wish to see the lectures printed persuaded me to collect them and offer them for publication.

The lectures were addressed to people who are curious about the physicist's world, and I attempted to describe it as I would to a close friend in another field. In other words, I tried to show not only the main discoveries but also the way they were reached, the personalities of the leading physicists, and the errors committed before the right path was found. The human side and the succession of events are often full of drama.

Experience has also shown me that many young scientists wish to know more about the personalities of important scientists rather than just names attached to some discovery, and I hope that this book may at least partly satisfy such legitimate curiosity.

The book does not pretend in any way to be a history of modern physics and even less to be a small physics text. It is, rather, an impressionistic view of the events as they appeared to me during my scientific career, which started about 1927. Naturally they cannot be treated out of context, and for this reason the story begins earlier. The choice of persons and subjects treated is thus subjective, limited, and colored by my personal experience.

I am indebted to the late Mrs. Laura Fermi, Professor J. Heilbron, and several of my contemporaries and colleagues for criticism and suggestions. I thank Professor F. Rasetti, l'Institut Solvay, CERN, California Institute of Technology, Lawrence Berkeley Laboratory, and others for illustrations.

January 1980   Emilio Segrè

FROM X-RAYS TO QUARKS

Modern Physicists and Their Discoveries

Chapter 1

Introduction

Mathematics and *physics* are words that often evoke unpleasant memories of concepts that were difficult to understand and appealed to odd individuals. When I went to school, teachers occasionally called science a "dry" subject, and many pupils agreed. The sight of mathematical formulae in print was a sure indication of incomprehensibility or even of black magic. Even now science is so often accused of nefarious doings that one forgets that it has also done some good.

Despite these negative views, scientific research is just as fascinating, dramatic, and full of human interest as artistic creation. However, most of the historical and biographical aspects that are given prominence in literary or artistic disciplines are often omitted in the teaching of science. This is probably because of the cumulative character of science. If there had been no Newton, someone else would have invented calculus and discovered gravitation, but without Shakespeare there would not have been *Hamlet*. It is thus considered more justifiable to study Shakespeare's life than Newton's.

I believe, however, that physics, too, has a rich human component, and it is mainly this element that I want to describe here. I limit myself to physics because it is the field in which I have direct knowledge. I hope this familiarity might enable me to communicate some of the inspiration, the creative effort, and the drama entailed in scientific work.

These historical aspects should interest not only physicists. It has often been said, probably correctly, that the nineteenth and twentieth centuries are an era as brilliant and as unique for science as the Renaissance was for art. He who has had the good fortune to be a contemporary of the Michelangelos or Shakespeares of this age can recall it with an immediacy and pathos that go beyond what can be grasped from the works alone. Although one of the major exponents of this renaissance, Marie Curie, said, "En science nous devons nous intéresser aux choses, non aux personnes" [In science we should concern ourselves with things, not people.], I believe that this judgment is too stern.

1

In this book I will try to evoke the personalities of some of the major physicists of this century and to point out some of their achievements, attempting to make them understandable to the layman. With some good will, and with some gaps, this is possible. I will avoid becoming so technical as to be intelligible only to the professional. Occasionally readers may skip a few pages if they find them too difficult, without losing the thread of the events.

Some knowledge of physics, however, is necessary. Whereas we can all look at Michelangelo's *David* or read *Hamlet* (and even here there are great differences in appreciation depending on our background), it is not possible to understand the double nature of light quanta or Schrödinger's equation without some preparation. Mathematical formulae simplify the tale. Mathematics is the natural language of physics, as Galileo pointed out, and although Volta and Faraday wrote great physics without using a formal mathematical language, they thought mathematically, and their ignorance of standard mathematics makes them less, not more, intelligible.

We must never forget that many scientific advances were achieved by the contributions of a multitude of workers, who prepared the terrain and did essential spade work. These people are often unknown or forgotten as individuals, but collectively they are indispensable. Furthermore, scientific events are interrelated and may overlap in time or space. If one tries to follow too closely, this intricate counterpoint can lead to complication and confusion. I have thus chosen to follow the trend of events, sometimes at the expense of strict chronological order.

The Physicists' World in 1895

It is natural to start our story around 1895 because in two or three years at that time physics took a decisive turn: A few experimental discoveries opened up microscopic consideration of the atomic world. Chemists had known of atoms for at least a hundred years, and through the kinetic theory of gases physicists had also made good use of atomic ideas, but nothing was known about atomic composition and structure.

In the Western world, where the knowledge of atomic structure began to unfold, England, France, and Germany were the three leaders in science. The three big powers were all experiencing different political and social situations. England was at the peak of her splendor under the rule of Victoria Regina and Imperatrix. The queen, who had become Empress of India in 1876, had been on the throne since 1837. The celebration of her jubilee in 1887 turned into a demonstration of the country's loyalty to her and pride in her empire. Enriched by 2,500,000 square miles of recently acquired territory, Britannia "ruled the waves" in splendid isolation.

France was still smarting from the defeats in the Franco-Prussian war of 1870 and 1871, which had been a tremendous blow to her ego and to the self-image of all French people. The demoralization of the French can be measured by the reaction of Pasteur and other French scientists to the disasters of the war. Distressed, wounded in their deep-rooted patriotism, they associated France's defeat with her neglect of science over the previous fifty years and recalled with pride the part played by science in the defense of the country during the Revolution and the Napoleonic wars. Pasteur hoped that through science he might be able to hasten France's recovery.

Germany, rapidly ascending and dominated by the military, had set off on an imperial course. The long struggle between civilian and military authority, which had lasted over sixty years, had been unfortunately resolved with the predominance of the military. Bismarck had been fired in 1890. Kaiser Wilhelm II (1859–1941) was young as rulers go and not experienced. Considering himself very intelligent—which he was not—he believed that he was ruling Germany superbly and bringing her to glorious times. At the start of the First World War he said, "Ich führe euch herrlichen Zeiten entgegen" [I lead you to glorious times"]. So much for his judgment.

In the world of 1895 there were no airplanes, virtually no telephones, and very little electricity. The ocean could be crossed on a steamship, but even then, seventy-five years after transatlantic boats had begun using steam, the steamship was occasionally equipped with supplementary sails. The main form of communication was mail, not only between distant places but also within cities. Paris, for instance, had a quite rapid system of pneumatic mail: a network of pipes in which letters were moved by compressed air. Streets were lit by gas.

In 1895 there were no automobiles; but two years later when Ernest Rutherford visited the exhibition at the Crystal Palace in London, he wrote to his mother, "The chief point of interest to me was the horseless carriage, two of which were practicing on the ground in front." They traveled at about 12 miles an hour but made "rather a noise and rattle." However, traffic accidents did happen even in the absence of automobiles, when horses drawing cabs or carts ran out of control; a few years later, in 1906, science was to lose one of its highest exponents to such an accident. There was no smog, but the streets reeked of manure, as inevitable a consequence of the transportation of those times as exhaust fumes are of our gasoline-propelled vehicles. Cities were smaller and more beautiful than at present, but they often had inadequate sanitation.

Physics laboratories were very different in organization and equipment from our present ones. There was usually only one professor, who often had his residence at the laboratory, and who was helped by very few assistants. Now when we rank the facilities of an institution, we do so according to the energy of its accelerator or perhaps the cooling capacity of its cryogenic establishment. But in 1895 accelerators and modern cryogenic plants were far in the future, although the liquefaction of air on a commercial scale had been achieved by that year.

One way of ranking a laboratory was according to the power of the battery it owned. Laboratories in those days needed electricity for experiments, but they could not draw electricity from the mains for the simple reason that mains seldom existed, so they kept batteries in their basements. A battery consisted of a collection of cells; the larger the number of cells, the higher the status of the establishment. Several types of cells had been developed since Volta's original "electric pile" of 1800. They were all based on the same principle but varied in the composition of their electrodes and in their electrolytic solutions. Many scientific laboratories used Bunsen cells, which could reach a high voltage (up to 1.95 volts) and could deliver heavy currents. But it was quite a task to keep them in working condition. They contained sulfuric acid and nitric acid, which corroded the zinc anode and exuded strong, objectionable fumes.

There is detailed advice for handling batteries of Bunsen cells in a French physics textbook by Adolphe Ganot, published in 1863. (It was the Italian edition of this book that introduced me to physics when I was about eleven years old.) In rereading it recently, I was impressed by the vivid advice, excerpts of which I translate below:

> The mixture of water and sulphuric acid must be prepared in advance. ... First pour the water into a wooden tub, then add one tenth in volume of common sulphuric acid, so that the solution will indicate 10 to 11 degrees on the Baumé acid scale. If a Baumé scale is not available, the water is sufficiently acidulated when it becomes lukewarm and a drop of it placed on the tongue cannot be held there. Cells must be placed ... on a very dry wooden table. ... Then with a funnel pour nitric acid in the porous inner container up to two centimeters from the top. ... The truncated cones that fit in the carbon must be carefully cleaned with sandpaper to ensure good connection. ... What must be watched above all is the amalgamation of the zinc plates. A plate needs to be amalgamated when a hissing sound is heard in the acidulated water while the pile is not in use ... the acidulated water may also steam and even boil. ... To amalgamate the zinc plates ... place them, one after another in an earthenware vase containing some acidulated water and two kilograms of mercury and spread it on the plates with an iron brush. ...

One of the important instruments of the time was the Ruhmkorff coil (induction coil), which was used to produce high potential differences and long sparks (Figure 1.1). The instrument consisted of two coils wound around a cylindrical iron core and insulated from each other. An electric battery produced a current in the primary coil, and this current was repeatedly interrupted by a breaker. The variation of the primary current induced a current in the secondary coil, creating a potential difference between the terminals of the secondary. Whereas the primary coil was made of thick wire with few turns, the secondary was a thin wire with so many turns that was miles long. A large Ruhmkorff coil of this period, preserved at the Royal Institution of London, has a secondary coil 280 miles long and could make sparks 42 inches long. So the length of sparks, like the power of a battery, could serve as a standard to rank a laboratory.

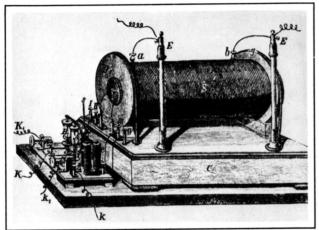

Figure 1.1 A Ruhmkorff coil. (From Urbanitzky, *Electricity,* 1890.) It is a transformer in which the current in the primary coil is suddenly broken. This generates a high voltage in the secondary coil, which gives off a spark in the air. The coil served to supply discharge tubes.

The production of vacuums has dominated physical research for over 100 years, and all the advances in investigation of the atom were coupled with advances in vacuum technology. In the laboratories of 1895 vacuums, created by primitive pumps, were needed for experiments on the discharge of electricity through gases, experiments that resulted in the discovery of x-rays and of the electron not long afterward.

Figure 1.2 shows the pump that was used by Sir William Crookes in his investigations of electrical discharges in vacuum tubes. The tubes to be evacuated were connected to the pump through the drying tube containing phosphoric acid, at the right. The mercury in the container at the left came down the fall tube drop by drop, driving the air out of the apparatus bubble by bubble. The level of the mercury in the gauge tube, compared with the level in a barometer, indicated the degree of vacuum obtained. The mercury reservoir had to be lifted and lowered manually many times, a strenuous task for the technician in charge of evacuating the professor's tubes and containers. In all such pumps the standard for the perfect vacuum was the barometer. The vacuum attainable with such a pump was about a million times worse than what we would call a decent vacuum today.

To learn in some detail what physicists were doing at the turn of the century, let us take a look in one of the leading journals of the time, the *Annalen der Physik.* A little earlier, the same journal was entitled *Analen der Physik und Chemie* because the sciences of physics and chemistry were still considered jointly, in contrast with our trend toward specialization, which has created a journal for each subbranch of physics. The subjects treated in the *Annalen* were the liquefaction of gases; the measurement of specific heats; electromagnetic waves, and especially attempts to reproduce with electromagnetic waves all the phenomena of optics; reflection and refraction; diffraction, rotation of the plane of polarization; and so on. Thermodynamics was then about forty years old and not yet entirely consoli-

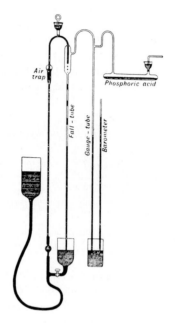

Figure 1.2 A mercury vacuum pump. (From S. P. Thompson, *Light Visible and Invisible*, 1897.) The pump works by trapping air in the fall tube. The gauge tube compares the vacuum obtained with that in the barometer.

dated. Gas discharges were studied with the Ruhmkorff coil and with tubes like the one in Figures 1.6 and 1.7. The kinetic theory of gases was developing vigorously, although not many people were interested in it, and some of the great figures in this field received little recognition. Josiah Willard Gibbs (1839–1903), teaching at Yale University, was ignored by most of the scientific world (except Maxwell and a few others). Ludwig Boltzmann (1844–1908), one of the founders of statistical mechanics, complained in Viennna that nobody in the German-speaking countries was paying attention to his work. Other topics treated in the journals of those times were physical chemistry and ionic dissociation, the beginning of the concept of ions in solution, and the relationship between thermodynamics and chemical equilibrium. No one thought seriously of making models of atoms; not only would this have been beyond feasibility, but the atom had not yet gained full recognition.

Of course, chemists knew of the atomic "hypothesis," but belief in the reality of atoms was not shared by all. In retrospect it would seem that since chemists wrote chemical formulae and were acquainted with Avogadro's law and Faraday's laws of electrolysis, they should also have believed in atoms. But this was by no means the case. As late as 1905 skepticism was still widespread, with some scientists rejecting outright the corpuscular theory of matter and others recognizing the usefulness of the atomic theory in chemistry but regarding it as remote from reality. These skeptics were neither crazy nor incompetent. For instance, somewhat earlier the Wayn-flete Professor of Chemistry at Oxford, Sir Benjamin Collins Brodie (1817–1880), wrote papers and books to show that atoms were not necessary to chemistry. In all earnestness he developed a system from which atoms were excluded, which he called "ideal chemistry." He was provoked when wires and balls were used for building models of molecules in organic chemistry; he considered these constructions a "thoroughly materialistic bit of joiner's work," an outrage, something absolutely beneath the dignity of chemistry.

In 1887 the banner of antiatomism was taken up by Wilhelm Ostwald (1853–1932), a prominent German chemist and an early winner of the Nobel Prize (1901). That year, as professor of chemistry at Leipzig, Ostwald delivered an inaugural address in which he advanced a doctrine of "energetics," claiming that all phenomena could be explained through the interplay of energy without the need for atoms. He later published a textbook of chemistry that did not use the atomic theory, and as late as 1909 this book was translated into English as *Fundamental Principles of Chemistry.* Ostwald kept his stand until J. J. Thomson and S. A. Arrhenius managed to shake his conviction, and he recanted in the 1912 edition of his *Allgemeine Chemie.*

Among physicists, one of the most notable nonbelievers in the atomic "hypothesis" was Ernst Mach (1838–1916), also a distinguished psychologist. In the 1906 edition of *The Analysis of Sensations* he spoke about "the artificial hypothetical atoms and molecules of physics and chemistry" and without denying "the value of these implements for their special limited purposes," he compared them to the symbols of algebra. Only after seeing the scintillations of alpha particles was he convinced that atoms did exist, or at least he relented in his skepticism.

The reason for the widespread skepticism was not so much contrariness as the fact that no one had ever "seen" an atom in a way that was convincing to the mind's eye. Even today no one has ever seen an atom in a literal sense, but the evidence for atoms is more convincing than the evidence for some things that many persons have "seen," such as miracles and flying saucers. We must also remember that although Avogadro's law—that equal volumes of gas at the same temperature and pressure contain the same number of molecules—was formulated in 1811, it was not until 1860, almost fifty years later, that Avogadro's number—the number of molecules in a mole—was "measured," or that scientists got an inkling of its magnitude, together with the order of magnitude of several atomic quantities, such as atomic size and mass.

At the end of the nineteenth century even a man such as Max Planck was cautious about giving credence to the atom. As he recalled in his *Scientific Autobiography,* he was "not only indifferent, but to a certain extent even hostile to the atomic theory." He accepted it only when it became necessary for the theoretical foundation of his radiation law.

Leading scientific personalities, could easily be identified, country by country. In the United Kingdom the list would include Lord Kelvin (William Thomson; 1824–1907). In 1895 he was 71 years old, had been a baron for three years, had held the chair of natural philosophy at the University of Glasgow for almost half a century, and was considered the leading physicist of the realm. His influence on the generations of students he trained and inspired either directly or through his writings had been enormous. An even greater contemporary of Lord Kelvin's, James Clerk Maxwell (1831–1879), had died young, and the realization that he was one of the greatest physicists who ever lived did not come until later. Other British luminaries of that period were Lord Rayleigh (1842–1919), the chemist Sir William Crookes (1832–1919), and Sir William Ramsay (1852–1916). In 1884 J. J. Thomson (1856–1940) succeeded Lord Rayleigh as Cavendish Professor at Cambridge University, a post he held for thirty-five years. Yet as generations of physicists go, he belonged to a younger group. Michael Faraday (1791–1867), who to us seems almost prehistoric, was no more removed from a man of 1895 than Max Planck, who died in 1947, is from us; in fact, he probably seemed closer, because in the last century science proceeded at a slower pace.

The French scientific scene was dominated by Louis Pasteur (1822–1895), biologist, chemist, and physicist, who died in the very year 1895. There was no French physicist of Pasteur's scientific stature; Ampère (1775–1836), Fresnel (1788–1827), and Carnot (1796–1832) were already figures of the past. Pasteur was the representative of French science, a tremendous scientist, a great benefactor of humanity, and an appealing personality—at least when seen from a distance. Although those who write about him depict him as a saint, the numerous polemical papers he wrote suggest that he must have been a quarrelsome man. He embodied the optimistic spirit of the end of the century. There was hope in progress; confidence that science would solve all problems and that scientists and other thinkers would ultimately inspire all people with ideas of concord and justice—a conviction that was destroyed by the First World War and has not returned since. Pasteur eloquently proclaimed that laboratories were the temples of humanity, that peace would prevail over war, and that science would lead to very great times.

The towering figure in German science was Hermann Ludwig Ferdinand von Helmholtz (1812–1894). Helmholtz held a unique, very influential position in Germany, similar to the position held by his friend Lord Kelvin in England. He was Maxwell's rival, and some of his work on the theory of electrodynamics, at variance with Maxwell's theory, kept scientists divided for several years. It was Helmholtz's most distinguished and beloved pupil, Heinrich Hertz (1857–1894) who resolved the conflict in favor of Maxwell. In 1887, Hertz performed the famous experiments in which he demonstrated electromagnetic waves, proving beyond any doubt the validity of Maxwell's equations and opening the field of radiocommunication.

Figure 1.3 Hendrik Antoon Lorentz (1853–1928). Lorentz's work marks the limits achieved by classical physics and forms the link between Maxwell's generation and that of Einstein and Planck. Lorentz's personality had a strong influence on the physicists' world because of the respect he commanded. (Nobel Foundation.)

These were the "grand old men" of 1895, but there was also a younger generation of scientists. It included J. J. Thomson in England, Ludwig Boltzmann in Austria, Max Planck (1858–1947) and Philipp Lenard (1862–1947) in Germany (Einstein was then only sixteen years old, an unknown, and a not too successful student), and Jules-Henri Poincaré (1854–1912) in France. Poincaré was just in his forties, yet he was already universally recognized as the greatest living mathematician. He was interested not only in mathematics but also in physics, astronomy, and philosophy, and was reckoned among French literary talents. His famous Sorbonne courses had been instrumental in opening the seals of Maxwell's *Treatise on Electricity* to continental scholars.

H. A. Lorentz (1853–1928) was also one of the luminaries of that time (see Figure 1.3). He represented the Netherlands, which enjoyed such an extraordinary scientific flowering that several of its best scientists had to emigrate for lack of positions at home.

I estimate that in 1895 there were approximately a thousand physicists, as compared with the sixty thousand there may be today. They were reasonably well paid and were held in moderately high regard. It is not true, as some people believe, that the vital importance of science has only recently been recognized. A scientist like Helmholtz could approach the Kaiser whenever he wanted; the Kaiser, demonstrating his interest in science, had the new Potsdamer Brücke decorated with the statues of C. F. Gauss, the "Prince of Mathematicians," W. C. Röntgen, H. Helmholtz, and the electrical inventor and industrialist W. v. Siemens. (The statues were destroyed during the Second World War.) In France Napoleon III, imitating his illustrious uncle, received important scientists at court. England knighted her successful scientists no less often at that time than now and occasionally raised the greatest to the peerage—witness Lord Kelvin. Lord Rayleigh was often visited at his estate by his political friends, including the prime minister.

These times, which saw the birth of a new physics, were characterized by fresh and adventurous thinking everywhere, not only in science. They were times of social and intellectual ferment, of exaltation of the individual in literature, and of revolt against academicism in the arts, with only architecture stubbornly proclaiming the virtues of the past. Socialist movements were afoot everywhere, while anarchism was reaching its climax of violence with the assassination of royalty and heads of state.

In France the Impressionists and other new schools of painting were at work, although their controversial paintings were not highly valued and those of Van Gogh could not be sold at all. In the world of music Debussy was still composing in semi-obscurity. The two great literary figures were Anatole France and Emile Zola. France portrayed the special character of his country in his own times, and Zola was a very popular naturalistic novelist who indicted French society when he supported Dreyfus's innocence in the "affair" that kept the country in turmoil for several years at the end of the century.

England had gone through the harrowing intellectual experience of assimilating Darwin's theory of evolution. In 1895 Darwin had been dead thirteen years, but Herbert Spencer, the philosopher of the scientific movement in the Darwin tradition, was very much alive at seventy-five, still thinking and writing, expressing the faith in progress that was peculiar to those times. In literature Oscar Wilde was delighting his audiences with his witty dramas, while Thomas Hardy and George Meredith, in a somber vein, were writing the last of the Victorian novels. H. G. Wells and G. B. Shaw were in the early stages of their careers. Shaw, still a beginner as a popular dramatist, was exerting a greater influence as the moving spirit of the socialist Fabian Society than as a writer.

Germany had produced Friedrich Nietzsche, the poet of *Thus Spake Zarathustra* and philosopher of the superman moved by the "will to power." Nietzsche's doctrines were a complete and drastic expression of the innovative spirit and the revolt against accepted values at the end of the nineteenth century. In 1895 Nietzsche was ailing and no longer writing, but his influence was beginning to be seen in the works of others; for instance, the poetry and novels of Gabriele D'Annunzio, an esthetic and fanatic exponent of Italian nationalism and future "superman" in World War I. Unfortunately, Nietzsche's ideals were to inspire more dangerous men: the future dictators Mussolini and Hitler.

The prevailing optimism and faith in science at the end of the century were expressed in the much acclaimed Ballet Excelsior, an Italian creation that was produced in Italy and France for at least three decades. The allegory showed Enlightenment leading Obscurantism, the genius of darkness, to witness the great discoveries and inventions of a civilization inspired by Divine Power: the steamship, the "electric telegraph," the Suez Canal, and the tunnel of Mont Cenis. Obscurantism then had a vision of all people united in brotherhood while under his own feet the earth caved in and engulfed him. The ballet closed with the triumph of Science, Progress, Brotherhood, and Love.

New Horizons

We now turn to physics proper and to the work that led to the understanding of atomic structure. The years from 1895 to 1897 were crucial because of four great discoveries: x-rays, the electron, the Zeeman effect, and radioactivity. Although the discovery of the electron was chronologically the last (it was completed only in 1897 with the measurement of the ratio of its charge to its mass), I shall deal with it first since it is part of the outcome of nineteenth-century physicists' long-standing concern with the discharge of electricity in "vacuum" tubes.

In an early study (1833) of electrical discharges in gases, Michael Faraday found that "rarefaction of the air wonderfully favors the glow phenomena" ("Experimental Researches on Electricity"). He examined the glow in several different gases at low pressure and was never able to separate it "into visible elementary intermittent discharges." He described the beauty of this glow and observed the dark space near the anode that now bears his name. The vacuum he could attain was very poor, poorer than he could realize, although he was aware that "as perfect a vacuum as could be made" was far from being an absolute vacuum.

In 1858 Julius Plücker had the idea of bringing a magnet near a "vacuum" tube to see what would happen to the discharge. Plücker (1801–1868) was a German mathematician, a topologist who later in his career became a professor of physics at Bonn and developed an interest in the relation between magnetism and gas discharges. (We must remember that it was quite normal at this time for a scientist to span more than one field: Gauss was both a mathematician and a physicist; Helmholtz was a physiologist, a physicist, and a philosopher; and Kirchhoff was a chemist as well as a professor of theoretical physics.) When Plücker brought a magnet near his vacuum tube, he noticed some deflection of the discharge. The next year he reported seeing a bright green phosphorescence in the glass of the tube near the cathode and was able to make the patches of phosphorescence change position by using a magnet. However, he could not go any further because his vacuum was too poor.

In 1869 Plücker's pupil Johann Hittorf (1824–1914) was more successful. In the intervening years the first mercury pumps had come into use, and he could evacuate his tubes a little better than his predecessors did. He saw the shadow cast by an object placed in front of the cathode, an indication that the discharge originated in the cathode. The name *Kathodenstrahlen*, cathode rays, was coined in 1876 by E. Goldstein (1859–1930). In 1879 William Crookes carried out systematic investigations of the cathode rays in tubes he evacuated with an improved pump that he had devised himself. It is illuminating to read in one of Crookes's papers that he was certain he had an especially good vacuum. The pressure in it was 40×10^{-3} millimeters of mercury, about a million times higher than what we have in modern large accelerators, and I wonder how well he measured it.

Today we know that cathode rays are swiftly moving electrons, but nobody then knew of even the existence of electrons. The known facts about cathode rays were that they came out of the cathode of a highly exhausted tube; hit the wall of the tube on the opposite side, making it luminous; traveled apparently in straight lines since objects on their path cast sharp shadows; and were probably deflected by a magnet, although no one could be positive about this.

A great dispute arose over the nature of cathode rays. What were they? Some said they were corpuscles, particles projected from the cathode; others believed they were waves. Strangely enough, opinions divided along national lines. In 1892 Hertz claimed experimental evidence that the cathode rays could not be particles and so they must be waves. Gustav Heinrich Wiedemann, (1826–1899) Goldstein, and all German physicists agreed. But in England Crookes insisted that these rays were electrically charged particles (Crookes called them "radiant matter"), and physicists in England—Kelvin, J. J. Thomson, and others—went along and supported "particles."

Finally, in 1895 in France, Jean Baptiste Perrin (1870–1942) found substantial proof that cathode rays were negatively charged particles. Having produced cathode rays in a well-evacuated discharge tube, he sent them into a Faraday cage and showed that they carried a negative charge. They could be deflected by a magnet and brought in or out of the Faraday cage, depending on how the magnet was moved. These were important experiments that opened the way to further progress.

Jean Perrin was a distinguished French physicist educated in Paris at the École Normale Supérieure. He was also the father of another important physicist, Francis Perrin, and thus was the founder of one of the prominent French scientific dynasties. (Other major scientific families were the Becquerel, the Curie, and the Brillouin families, some of whom we will consider later.) Before the First World War Jean Perrin did elegant experiments on Brownian motion using colloidal spheres of gutta-percha, which acted as giant molecules. From these experiments he indirectly determined the charge of the electron. In later life he participated actively in French politics as a militant leftist. He died in New York, having fled from occupied France.

Between the work of Perrin and that, directly related, of J. J. Thomson, some important work was done on the *bound* electron with spectroscopic techniques; however, it was pertinent to the general problem of the electron. All these activities overlap in time, and there is no doubt that the various scientists read the scientific journals and were informed of the work in different laboratories. We now turn to Holland and Pieter Zeeman.

Pieter Zeeman

The name *electron* had already been suggested by G. Johnstone Stoney in 1894. Moreover, It was believed that electric charges moving within the atom were responsible for light emission. Several other phenomena supported the idea of point electric charges, but views on the subject were vague. Suddenly in 1894, Pieter Zeeman (1865–1943), an unknown young physicist working at Leyden, made a substantial discovery. His findings were quickly followed by a theoretical explanation provided by the already famous H. A. Lorentz.

Zeeman was born in the Netherlands. He studied under Kamerlingh Onnes and later became an assistant to Lorentz. Zeeman still read Faraday as living science and as a source of inspiration. He noted that Faraday, in his constant quest for a connection between the various "forces of nature," had tried to influence light by magnetism. These attempts had brought the important discovery of the magnetic rotation of the plane of polarization of light produced in glass by a magnetic field; there is a famous portrait of Faraday with a piece of flint glass in his hand that immortalizes the event. In 1862, in one of his last experiments, Faraday had also unsuccessfully tried to influence the light emission of sodium vapor by a magnetic field. Maxwell, in 1870, had also denied the possibility of this phenomenon.

Zeeman realized that he had much better apparatus than Faraday. A small effect could have escaped Faraday, who worked with prism spectroscopes of low resolving power, whereas Zeeman could use a diffraction grating. Announcing the discovery of the "Zeeman effect," he said:

> If a Faraday thought of the possibility of the above mentioned relation, perhaps it might yet be worthwhile to try the experiment again with the excellent contrivances of spectroscopy of the present time, as, so far as I know, this has not been done by others as yet. [*Philosophical Magazine* [5] *43*, 226 (1897)]

He tried the experiment and soon observed a slight broadening of the spectral lines brought about by the magnetic field (see Figure 1.4). He noted that the edges of the broadened lines were polarized, and by refining the technique, he saw a triplet or a doublet according to the relative orientation of the direction of the observation and of the magnetic field. He communicated his discovery to Lorentz, who immediately gave an explanation of the observations.

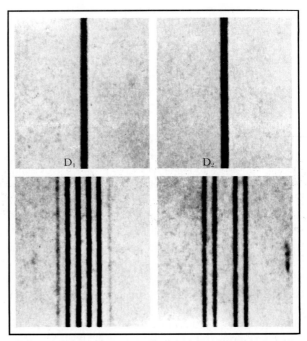

Figure 1.4 The Zeeman effect on the yellow spectral lines (D lines) of sodium. Above, lines D_1 and D_2 without a magnetic field. Below, the same lines are split into multiplets when the source is placed in a magnetic field. The multiplet lines are polarized. The direction of observation is at right angles to the direction of the magnetic field.

The fundamental idea was that light was emitted by charged particles (electrons) moving in the atom. Their motion was influenced by the magnetic field according to the classical laws of electromagnetism. From the change in frequency of the emitted light, Zeeman and Lorentz were able to determine e/m, the specific charge of the particles causing the light emission as well as the sign of the charge. Zeeman first made a mistake in the sign, which he soon corrected; in a matter of sign even such a meticulous Dutchman as Zeeman can go astray.

Zeeman's most remarkable findings were the negative sign and the value of e/m, which was about a thousand times larger than one expected for an entire atom on the basis of the primitive notions scientists then had about atomic masses. The Zeeman effect later proved to be a powerful tool for unraveling atomic structure, and decisive for the discovery of Pauli's principle, for the electron spin, for details on the mechanism of emission, and more. It fit perfectly with quantum mechanics and became important experimental evidence for it.

In Zeeman's experiments the electrons were bound within atoms. At about the same time free electrons also made their entry into physics, mainly through the work of J. J. Thomson.

Joseph John Thomson

At the time the electron was discovered, J. J. Thomson was the Cavendish Professor at Cambridge University (see Figure 1.5). Born in 1856 near Manchester into a family of business people, he was expected to follow the family tradition, but circumstances led him toward scientific studies, and in 1876 he was admitted to Trinity College, Cambridge. Here he trained himself, as was usual at the time, for a stiff competitive written examination called tripo from the tripods full of charcoal used to heat the rooms in which the competition was held. The winners were called wranglers. In 1880 he came out as second wrangler, as Maxwell had before him.

J. J. Thomson heard some of Maxwell's lectures, but it was under Maxwell's successor as Cavendish Professor, Lord Rayleigh, that Thomson completed several theoretical papers. In 1884 Rayleigh resigned as Cavendish Professor according to his original commitment, in which he had promised to serve for five years. Thomson applied for the professorship, he relates, "without serious consideration of the work and responsibility involved." He was only twenty-eight years old and did not expect to be selected. To his surprise, he was. The electors were either very lucky or very far-sighted. Thomson says, "I felt like a fisherman who with light tackle had casually cast a line in an unlikely spot and hooked a fish much too heavy for him to land. I felt the difficulty of following a man of Lord Rayleigh's eminence." It is remarkable that he does not mention Maxwell, although in another place Thomson says about the appointment of the first Cavendish Professor (February 1871),

> It is believed that the University had first approached Sir William Thomson (later Lord Kelvin) and then von Helmholtz, the great German physicist and physiologist, but neither of these could see his way to accept the post. At the time of his election Maxwell's work was known to very few, and his reputation not comparable with what it is now (1936). ... Indeed even at the time of his death the truth of his supreme contribution to physics—the theory of the electromagnetic field—was an open question. [*Recollections and Reflections*, pp. 96, 101]

Figure 1.5 Joseph John Thomson (1856–1940), the famous English physicist, celebrated for his experiments on the electron and on isotopes. He was the third director of the Cavendish Laboratory. A portrait of him studying a cathode ray tube hangs in the Laboratory's Maxwell Lecture Room. Apparently he was clumsy but very quick in understanding the workings of apparatus. (Cavendish Laboratory, Cambridge University.)

Thomson proceeded to renew the laboratory, to introduce new methods of teaching, and to found a research school that was most successful. A stream of discoveries poured out of the Cavendish Laboratory: the electron, the cloud chamber, early important work on radioactivity, and isotopes are among the highlights. Among the pupils were Rutherford, C. T. R. Wilson, R. J. Strutt (Lord Rayleigh's son), J. S. E. Townsend, C. G. Barkla, O. W. Richardson, F. W. Aston, G. I. Taylor, and G. P. Thomson, all of whom became famous.

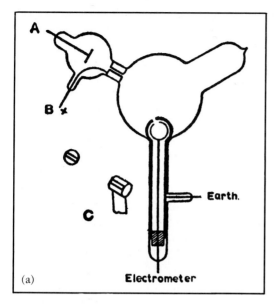

(a)

A

B

C

Earth.

Electrometer

Figure 1.6 (a) One of Thomson's discharge tubes from a drawing in *Philosophical Magazine* [44, 293 (1897)]. The electrons generated by cathode A can be deflected by an external magnet and brought to a collector (Faraday cage) connected to an electrometer that measures the total charge. (b) Another of Thomson's discharge tubes, from the same journal. A beam of cathode rays emitted by the cathode C and focused in A and B passes between D and E, where there is an electric field. A magnetic field perpendicular to the electric field is created by coils located outside the tube.

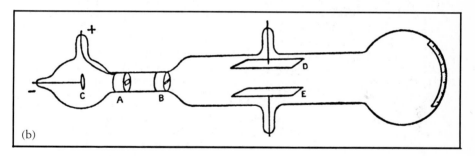

(b)

Röntgen's discovery of x-rays (discussed later in this chapter) gave a new method of ionizing gases and afforded a new insight into the behavior of gaseous ions. Thomson started work in that direction, and this led to the study of free electrons.

In 1897 Thomson confirmed the corpuscular nature of cathode rays and measured the velocity and the ratio of charge to mass of the corpuscles. Figure 1.6 shows two of the tubes used by Thomson in his experiment. In Figure 1.6a the rays were produced from the cathode A in the tube to the left. Through a slit in the anode B the rays passed into the second tube, and could be deviated using a magnet into a sort of Faraday cage. The charge collected was negative. It was thus demonstrated that cathode rays were negatively charged particles. Similar experiments had been done by J. Perrin in France. In a second type of tube (Figure 1.6b) the cathode rays generated at C were passed through the grounded slits A and B, forming a narrow beam of rays that traveled to the end of the tube. The place where the beam hit the bulblike end of the tube was marked by a small patch of bright phosphorescence.

When Thomson connected the two metal plates E and D (see Figure 1.6b) with the terminals of a battery, the phosphorescent patch moved, showing that the cathode rays were deflected by the electric field. With a magnetic field perpendicular to the electric field he could then deflect the rays magnetically. Magnetic deflection had been observed before, but J. J. Thomson was the first to observe the electric deflection. Its apparent absence was the first motivation for J. J. Thomson's investigation. Why had nobody seen electric deflection in the several decades during which cathode rays had been under investigation? The reason is simple: Unless there is a good vacuum in the cathode tube, no electric field can be established. A poor vacuum is a conductor, and no static electric field can be maintained in one. But Thomson succeeded, not only with the arrangement shown in Figure 1.6 but also with two others.

In August 1897 he wrote his now famous paper in which he described experiments "to test the electrified-particle theory" and applied the results of his measurements to the determination of the ratio of charge to mass of the particles making up the cathode rays; from the same experiments he derived also the velocity of the particles. Here is a summary of his reasoning. The total amount of electricity Q carried by a given current is equal to the number of particles N, or corpuscles, in it, times the charge e of each one:

$$Ne = Q$$

Then he measured the energy W transported by the corpuscles by measuring the heat developed; this had to be equal to the kinetic energy of the particles of mass m and velocity v

$$\tfrac{1}{2}Nmv^2 = W$$

Having deflected the beam of particles magnetically, he knew that

$$\frac{mv}{e} = B\rho$$

where ρ is the radius of curvature of the trajectory and B the magnetic field. Since the energy, the quantity of electricity, the magnetic field, and the radius of curvature were measurable, he was able to deduce that

$$\frac{e}{m} = \frac{2W}{Q^2B^2\rho^2}$$

had the value of 2.3×10^{17} esu/g, much larger than e/m for ions in electrolysis.

In his paper of 1897 Thomson made another remarkable observation: The corpuscles that constituted the cathode rays were the *same* no matter what the composition of the cathode or the anticathode or the gas in the tube. Here was a universal component of all matter.

A little later, in 1899, using techniques and ideas developed by his former pupil C. T. R. Wilson, J. J. Thomson measured the charge and the mass of the electron separately. Wilson had remarked that under favorable circumstances electric charges act as condensation nuclei for supersaturated vapors. They favor the formation of fog because water condenses on them. In such a fog, determined by the presence of electric charges, one can measure the volume of the droplets from the speed with which they fall, and their number from the total water precipitated or from the initial supersaturation. From this data one obtains the number of droplets contained in the fog. From the total charge transported by the fog, directly measurable, one then finds the charge on an average droplet, identical to the electron charge.

This work, done at the Cavendish Laboratory, gave the charge of the electron as about 3×10^{-10} absolute electrostatic units. From the measured value of e/m one also found the electron mass.

The drop method was later refined by R. A. Millikan (1910) in the United States. He observed not a fog, but single droplets, and he transformed the method into a precision one that gave the number 4.78×10^{-10} esu for the charge of the electron. For many years this value was the best directly measured. However, in 1929, to everybody's surprise, it turned out to be in error by about 1 percent, much more than the estimated possible error. The origin of the discrepancy was found to lie in a defective measurement of the viscosity of air. Today the value of the charge of the electron is known with the precision of about 3 parts per million; and it is 4.803242×10^{-10} esu; e/m, known with a precision of about 6 parts per million, is 5.272764×10^{-17} esu/g.

The discovery of the electron, important as it was, was overshadowed by another discovery that had taken place at the end of 1895. This great discovery was made by W. C. Röntgen (1845–1923), who stunned the world with his announcement of "a new kind of rays" and his demonstrations of what his rays could do.

Wilhelm Conrad Röntgen

Wilhelm Conrad Röntgen (Figure 1.7) was born in Lennep, a town in the Rhineland. His mother was Dutch, and his family moved to Apeldoorn in Holland when he was three years old. After studying in Holland, Röntgen went in 1865 to Zurich, where he enrolled in mechanical engineering at the

Figure 1.7 Wilhelm Conrad Röntgen (1845–1923) at the time of his discovery of x-rays. (Deutsches Museum, Munich.)

Polytechnic Institute. He studied first with Rudolf Clausius, the great thermodynamicist, and then with August Kundt, to whom he became very close. Kundt's most important contributions were in acoustics, but he is also known for an ingenious, albeit crude, determination of Avogadro's number.

Röntgen graduated from the Polytechnic Institute in 1868, and in 1869 he received a doctorate from the University of Zurich. In 1870 he returned to Germany as Kundt's assistant, first at Würzburg and later at Strasbourg; he became a Privat Dozent, and in 1875, having been appointed a professor of physics at a small German university, he started upon a normal academic career as a good, but not extraordinary, physicist. In 1888 he did an important piece of work showing that the convection current was the same as the conduction current. Such a finding may seem trivial today, but we must remember that Faraday worked hard to convince himself that the electricity from a cell was the same as the electricity produced by an electrostatic machine, a fact that was by no means obvious at his time. Röntgen proved that the current obtained by moving charges was the same as the current in a wire. He also made good measurements of specific heats, a standard field of interest in his day. He moved from one university to another, and in the fall of 1888 assumed his fourth post, a chair at the same University of Würzburg where he had been Kundt's assistant. It was a good

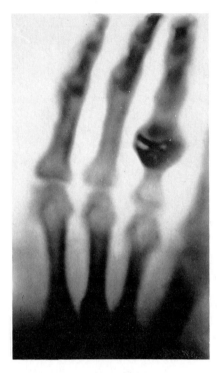

Figure 1.8 One of the very first images obtained by Röntgen using x-rays, or "Röntgen rays," showing the bones of a hand. He took the image on December 22, 1895. (Deutsches Museum, Munich.)

university, though not one at the very top. By the beginning of November 1895 Röntgen had written forty-eight papers now practically forgotten. With his forty-ninth he struck gold.

On the evening of November 8, 1895, Röntgen was operating a Hittorf tube and had covered it entirely with black cardboard. The room was completely darkened. At some distance from the tube there was a sheet of paper, used as a screen, treated with barium platinum-cyanide. To his surprise, Röntgen saw it fluoresce, emitting light. Something must have hit the screen if it reacted by emitting light. Röntgen's tube, however, was enclosed in black cardboard, and no light or cathode rays could come out of it. Surprised and puzzled by the unexpected phenomenon, he decided to investigate it further. He turned the screen so that the side without barium platinum-cyanide faced the tube; still the screen fluoresced. He moved the screen further away from the tube, and the fluorescence persisted. Then he placed several objects between the tube and the screen, and all appeared to be transparent. When he slipped his hand in front of the tube, he saw his bones on the screen (Figure 1.8). He had found "a new kind of rays," as he termed them in his first publication on the subject.

Röntgen was working all alone in his laboratory. He went on working by himself in the days that followed, not mentioning his observations to anyone. His wife noticed his preoccupation but did not know its cause and began to worry. All her husband would tell her was that he was working on something important. Later, he explained his reticence: He had been so astonished at his discovery, so incredulous, that he had felt the need to convince himself over and over again of the existence of these new rays. Finally he fixed his findings on photographic plates and at last became certain of his discovery.

On December 28, 1895, he handed a preliminary paper to the secretary of the Physical-Medical Society of Würzburg. It was printed at once, and by early January 1896 it was being distributed. In this sober, concise communication, Röntgen revealed none of his early feelings and doubts. The paper starts thus:

> If we pass the discharge from a large Ruhmkorff coil through a Hittorf or a sufficiently exhausted Lenard, Crookes, or similar apparatus and cover the tube with a somewhat closely-fitting mantle of thin black cardboard, we observe in a completely darkened room that a paper screen covered with barium-platino-cyanide lights up brilliantly and fluoresces equally well whether the treated side or the other be turned towards the discharge tube. ["Über eine neue Art von Strahlen," *Sitzungsberichte der Phys. Mediz. Gesellschaft zu Würzburg 137,* 132 (1895). Translated in *Nature 53,* 274 (1896)]

Röntgen went on describing the findings from his seven weeks of "secret" investigations: Objects were transparent to his "new rays" in different degrees; photographic plates were sensitive to x-rays; he could not observe any appreciable reflection or refraction of the rays, nor could he deflect them with a magnetic field; the x-rays originated in the area of the discharge tube where the cathode rays impinge on the wall of the glass tube.

On January 1 he had sent out the preprints, causing a great stir. His paper was unbelievable—but with it he also sent x-ray photographs of hands, which provided evidence that could not easily be dismissed. Among those who received this early communication were Boltzmann, Warburg, Kohlrausch, Lord Kelvin, Stokes, and Poincaré. Upon reading Röntgen's paper, many scientists ran to their laboratories, brought out their spark coils, and set about finding out whether they could see the x-rays. They did.

By January 1896, news of the discovery of x-rays had created a tremendous commotion all over the world. We can imagine the immense wonder over these rays to which almost anything was transparent and by means of which one could see one's own bones. Fingers without flesh but with the rings being worn were quite visible, as was a bullet lodged in the body. The implications for medicine were grasped immediately. One did not need to be a scientist to appreciate the discovery. The future physicist A. N. da Costa Andrade, a small boy in 1896, had been told that God could see everything and everywhere, and after hearing about x-rays, he began believing what he had doubted before. On January 23, Röntgen gave his only public lecture on his discovery, before the Physical-Medical Society. He was greeted with a storm of applause.

Some of this excitement reached me thirty years later when I was a student in Rome. An old curator of the Physics Institute, Augusto Zanchi, told me how once he had had to stay up all night to try to evacuate a tube with a mercury pump because the Queen of Italy wished to see the x-rays and had asked the professor of physics to give her a demonstration. Throughout the night the poor curator was afraid that he might not be able to evacuate the tube sufficiently, but fortunately he succeeded. The Queen had her show.

Röntgen's work on x-rays was perfectly competent in light of the knowledge of his times. But he could not gain an understanding of the nature of x-rays. In concluding his famous paper of 1895 he wrote:

> May not the new rays be due to longitudinal vibrations in the ether? I must admit that I have put more and more faith in this idea in the course of my research, and it now behooves me therefore to announce my suspicion, although I know well that this explanation requires further corroboration.

The "further corroboration" never came, but it took a good sixteen years and the work of Max von Laue and Friedrich and Knipping to settle the argument over the nature of x-rays.

In the months following his discovery Röntgen was invited to lecture all over the world, but he declined all invitations except one, for he wanted to continue to study his x-rays. He wrote brief notes to colleagues who asked him to demonstrate the new rays, expressing his regrets and stating that he had no time to give any lectures or demonstrations. The exception was the Kaiser, to whom Röntgen demonstrated his x-rays on January 13, 1896.

The idea of demonstrating to the Kaiser had kept Röntgen anxious. "I hope I shall have 'Kaiser luck' with this tube," he said, "for these tubes are very sensitive and are often destroyed ... and it takes about four days to evacuate a new one." But nothing got broken. An invitation at court such as the one Röntgen received involved, besides the lecture and demonstration, dining with the Kaiser, receiving a decoration (Kronen-Orden 2. Klasse), and, at leave-taking, walking backward out of respect for His Majesty. In this connection Richard Willstätter, the great organic chemist who unraveled the complexities of chlorophyll, tells that he and Fritz Haber, the synthesizer of ammonia, having made their discoveries, expected an invitation from the Kaiser. So they decided to practice walking backward. Willstätter was a collector of fine china, and a precious vase was in the room where they practiced. As might be expected, the exercises ended with the shattering of the vase. The two were not invited by the Kaiser, but their practicing was not a lost cause. Both later received the Nobel Prize and, according to protocol, had to walk backward after receiving the award from the hands of the King of Sweden.

On February 8, 1896, Röntgen wrote the following letter to his close friend Ludwig Zehnder, vividly describing the events surrounding his discovery.

Dear Zehnder! The good friends come last. That is the way it goes. But you are the first to receive an answer. Have many thanks for everything you wrote me. I can not as yet use your speculations on the nature of the X rays, since it does not seem permissible or advantageous to attempt to explain a phenomenon of unknown nature with a not entirely unobjectionable hypothesis. Of what nature the rays are is not entirely clear to me. And whether they are actually longitudinal light rays is to me of secondary importance. The facts are the important thing. In this respect my work has received recognition from many quarters. Boltzmann, Warburg, Kohlrausch, and (not least) Lord Kelvin, Stokes, Poincaré and others have expressed to me their joy over the discovery and their recognition. That is really worth a great deal to me, and I let the envious quietly chatter. I am not concerned about that at all!

I had not spoken to anyone about my work. To my wife I mentioned merely that I was doing something of which people, when they found out about it, would say, *"Der Röntgen ist wohl verrückt geworden."* [Röntgen has really gone crazy.] On the first of January I mailed the reprints, and then the devil was to pay! The *Wiener Presse* was the first to blow *"die Reklametrompete"* [the advertising trumpet], and the others followed. In a few days I was disgusted with the whole thing. I could not recognize my own work in the reports any more. Photography was for me the means to the end, but it was made the most important thing. Gradually I became accustomed to the uproar, but the storm cost time. For exactly four full weeks I have been unable to make a single experiment! Other people could work, but not I. You have no idea how upset things were here.

Enclosed I send the promised photographs. If you wish to show them in lectures, that is all right with me. But I would recommend that you place them under glass and frame, otherwise they will be stolen. I think that with the aid of the explanations you will have no difficulty with them; otherwise write to me.

I use a large Ruhmkorff 50/20 centimeter with a Deprez interrupter, and about 20 amperes primary current. My apparatus, which remains on the Raps pump, requires several days for evacuation. The best results are obtained when the spark gap of a parallel connected discharger is about three centimeters.

In time all apparatus will be punctured (with the exception of one). Any method of producing cathode rays will be successful, also with incandescent lamps according to Tesla and with tubes without electrodes. For the photography I use three to ten minutes, depending on the conditions of the experiments.

For your lecture I send you the compass box, the wood roll, the weight set, and the zinc sheet, as well as a nicely preserved photograph of a hand by Pernet of Zurich. But please return these items as soon as possible, insured. Do you have a larger screen with platinocyanide? Best regards from home to home.   Your Röntgen

[*Dr. W. C. Röntgen,* p. 87]

After Röntgen's discovery physicists and medical people rushed to investigate the new rays. In the year 1896 there were already more than 1,000 papers on the subject. Röntgen himself wrote only two more papers on x-rays in 1896 and 1897. He then returned to his old subjects, and in the next twenty-four years wrote seven papers of fleeting interest, leaving the investigation of x-rays to other younger, fresher forces. One can only speculate on the reason for this.

In 1902 Röntgen received the first Nobel Prize for physics. In 1900 he had moved to Munich, where he became director of the Institute for Experimental Physics. In 1914 he signed the famous manifesto of German scientists expressing solidarity with a militarist Germany, but he later regretted it. He suffered considerably during the First World War and in the following inflation, and died in Munich on February 10, 1923, at the age of 78.

Chapter 2

H. Becquerel, the Curies, and the Discovery of Radioactivity

The years around 1895 mark a turning point in physics not only because of the discovery of x-rays, the electron, and the Zeeman effect, but also because of the more revolutionary discovery of radioactivity.

At the end of 1895 Röntgen sent preprints of the announcement of his "new rays" to several colleagues including Henri Poincaré, the mathematician who had always shown great interest in basic physics research. Poincaré had participated in the debates over the nature of cathode rays, attempting to demonstrate their corpuscular constitution. Röntgen's discovery excited Poincaré possibly more than any other French scientist (although, in the first half of 1896, 135 communications and brief notes were published in the *Comptes-rendus* of the Académie des Sciences, alone).

Poincaré was a member of the Académie des Sciences and was in the habit of attending its weekly sessions. At the session of January 20, 1896, he showed the first x-ray photographs sent by Röntgen. When one of his colleagues at the Académie, Henri Becquerel, asked him from what part of the tube the rays emerged, Poincaré replied that the rays seemed to be emitted from the region of the tube opposite the cathode, the region where the glass became fluorescent. Becquerel immediately thought of a possible relation between x-rays and fluorescence, and the very next day he began testing whether fluorescent substances emitted x-rays, thus initiating a series of experiments which, within a few weeks, led to the discovery of radioactivity.

Becquerel's "Predestined" Discovery

Interest in phosphorescence and fluorescence ran in the Becquerel family. Henri Becquerel's father, Edmond, was the son of Antoine César Becquerel and Henri Becquerel was the father of Jean Becquerel. These representatives of four generations were all distinguished physicists, so for almost eighty years, from 1828 to 1908, there was always one, sometimes two, Becquerel in the Académie. It is worth saying a few words about this remarkable dynasty, partly because the story shows how Henri was predestined to discover radioactivity—he said so himself.

It is with Henri's grandfather that we must begin. Antoine César Becquerel (1788–1878) was born just in time to serve, as a young man, under Napoleon. He was one of the first officers to graduate from the Ecole Polytechnique, which over the years has always provided a rich source of technical, scientific, and military expertise for France. Many of the great men of French science were trained in this school, which put its students in military uniform, obliterating any mark of social class, and imparted strict discipline as well as a deep patriotic spirit. I once had occasion to lecture at the Ecole Polytechnique, and I remember reading above the dais from which I spoke the names of past professors: Arago, Ampère, Poisson, Fourier, Cauchy, Fresnel, Monge, Becquerel, and several others. The list was enough to intimidate any outside lecturer. It became a virtual requirement for a scientific, technical, or military career in France to be a *polytechnicien,* that is, a graduate of the Ecole Polytechnique, and Pierre Curie had difficulties because he was not.

The *polytechnicien* Antoine César Becquerel participated in the Napoleonic War in Spain in 1810–1812, but in 1815, after the fall of Napoleon, he resigned from the army. He had been wounded in the campaigns and was told that he was in poor health and had only a short time to live. (He lived to be ninety.) He then turned his attention to physics and soon became professor of physics at the Musée d'Histoire Naturelle in Paris and later its director. He wrote 529 papers and six textbooks, one of them in seven volumes. He did research on phosphorescence and dealt at length with phosphorescence in two of his books. He is known also for his studies on electricity and electrochemistry, in which he discovered some of the electrothermic effects. Altogether he was a well-known and much respected man. I remember often encountering his name when I read, as a boy, the old edition of Ganot's textbook that introduced me to physics.

Antoine César's son Edmond (1820–1891) followed very much in his father's footsteps, except for military service. He was admitted at the Ecole Polytechnique, became his father's assistant as the Musée, and later was a professor there. The professorship at the Musée was becoming a sort of hereditary trust, passed from father to son, and eventually to Antoine César's great-grandson Jean (1878–1953). Four generations of Becquerels lived, as Jean wrote in an article, in the *"même maison, même jardin, même laboratoire"* [the same house, the same garden, the same laboratory] in the Jardin des Plantes, in front of Cuvier's house. Edmond Becquerel studied the chemical action of light and was one of the first to photograph the solar spectrum. He was also a great expert in fluorescence, and the substance he

Figure 2.1 Henri Becquerel (1852–1908). Three generations of Becquerels, all physicists, were members of the French Academy, at times together. (Ciccione, Rapho.)

knew particularly well was uranium. He devised a fluoroscope and measured the intensity and duration of the fluorescence of uranium under the action of different lights.

By the time of Röntgen's discovery Edmond Becquerel's son Henri (1852–1908) had succeeded his father in the chair at the Musée d'Histoire Naturelle, had been named a professor at the Ecole Polytechnique, and had published papers on phosphorescence and fluorescence (Figure 2.1). It is now clear why, upon hearing from Poincaré of the discovery of x-rays, he thought that the two phenomena might be related. However, Henri's first experiments gave negative results: The phosphorescent or fluorescent substances he tested did not emit x-rays. Meanwhile, on January 30, 1895, the *Revue générale des Sciences* carried an article on x-rays by Poincaré in which he posed the question, "Do all bodies whose fluorescence is sufficiently intense emit both luminous rays and also Röntgen's x-rays, *whatever the cause of their fluorescence?*" If so, he stated, phenomena of this kind would not be associated with an electric cause.

Becquerel resumed his experiments, and this time he tried a uranium salt, uranyl potassium sulfate, which had previously been studied by his father. On February 24 he reported to the Académie:

> I wrapped a ... photographic plate ... with two sheets of thick black paper, so thick that the plate did not become fogged by exposure to the sun for a whole day. I placed on the paper a layer of the phosphorescent substance, and exposed the whole thing to the sun for several hours. When I developed the photographic plate I saw the silhouette of the phosphorescent substance in black on the negative. ... The same experiment can be tried with a thin sheet of glass placed between the phosphorescent substance and the paper, which excludes the possibility of a chemical action resulting from vapors that might

28

emanate from the substance when heated by the sun's rays. We may therefore conclude from these experiments that the phosphorescent substance in question emits radiations which penetrate paper that is opaque to light. ... [*Comptes-rendus de l'Académie des Sciences, Paris 122, 420 (1896)*]

It seemed as if x-rays were actually emitted by the uranium compound while it fluoresced. But a week later, when the Académie convened again on March 2, Becquerel knew better. The reason he knew better was that the weather in Paris had changed! He had tried to repeat the preceding experiments, but on February 26 and 27 the weather was poor and the sun was not out for sufficiently long periods. So he put everything in a dark drawer, leaving the samples of uranium salt in place, over the wrapped plates. He related to the Académie:

Since the sun did not show itself again for several days, I developed the photographic plates on the 1st of March, expecting to find the images very feeble. On the contrary, the silhouettes appeared with great intensity. I thought at once that the action might be able to go on in the dark. [*Comptes-rendus 126, 1086 (1896)*]

Figure 2.2 shows one of the "silhouettes" that Becquerel saw on developing his plates. He realized immediately that he had discovered something very important: The uranium salt emitted rays capable of penetrating black paper, whether or not it had been previously exposed to sunlight.

Here is a typical case of serendipity, in which chance, sagacity, and alertness were crucial ingredients. Henri Becquerel said also that his father and grandfather deserved credit for his discovery. According to him the work pursued for about sixty years in their laboratories was fatally bound, at the proper time, to lead to the discovery of radioactivity. But, at the time, Becquerel's discovery did not seem to measure up to Röntgen's and did not arouse the excitement that Röntgen's had. Scientists went on talking about x-rays and doing work on them, and left "Becquerel's rays" very much to their discoverer. By March 9, 1896, Becquerel had found that not only did the radiation emitted by uranium blacken the protected photographic plates, but it also ionized gases, making them conductors. From then on it was possible to measure the "activity" of a sample simply by measuring the ionization that it produced. The instrument used for this measurement was a crude gold-leaf electroscope.

The Curies and a Great Leap Forward

At this point in our story, about two years after Becquerel's discovery, Pierre and Marie Curie entered the scene. It is not that Becquerel did not continue his work; on the contrary, he pursued it with alacrity. However, he

Figure 2.2 The first plate marked by "Becquerel rays." It was placed under some uranyl potassium sulfate on February 26, 1896; after Becquerel developed it on March 1, he realized uranium emitted a radiation of unknown nature that did not depend on the phosphorescence of the uranium salt, as had been believed. The discovery of "Becquerel rays" was published in the *Comptes-rendus de l'Académie des Sciences de Paris* [*122,* 501 (1896)]. (CEA.)

restricted himself to uranium as the source of his rays; as uranium was the substance he knew best. Years later he wrote, "But since the new rays had been recognized with uranium, it seemed improbable, a priori, that the activity of other known bodies might be considerably bigger, and the research on the generality of the new phenomenon then seemed less urgent that the physical study of its nature." And so it was not Becquerel but the Curies who took the great step forward: They investigated other elements, and, by discovering first polonium and then radium, they prepared powerful sources that completely revolutionized the new science of radioactivity.

Madame Curie (1867–1934) née Marya Sklodowska, was born on November 7, 1867, in Warsaw (Figure 2.3). On her mother's side the family belonged to the minor nobility. Her father, Vladislav Sklodowski, a cultured gentleman who had studied in St. Petersburg, was a teacher of mathematics and physics; shortly after Marie's birth he became professor and assistant-inspector at a gymnasium, a kind of high school. Madame Sklodowska was the principal of a school for girls, a position she kept until Marie was several months old. She was a pious Catholic and raised her children accordingly; but soon after her death her teenage daughter Marie lost her faith and never returned to it. The family included a son, Josef, and three daughters, Hela, Bronya, and Marie.

The family was intensely patriotic, to the extent that the children participated in clandestine cultural activities in Russian-governed territory at the risk of imprisonment or worse. Throughout her history Poland had often been oppressed by her neighbors, and now she was oppressed chiefly by Russia. Polish schools were forced to use the Russian language and to adopt books that they did not want. The antagonism of the Polish population toward the dominant Russians was based on religious, political, and linguistic grounds. Feelings were strong, and the situation was tense. Marie was deeply shocked when one of her brother's friends was hanged for political reasons. In reaction to this oppression the Poles became fanatic patriots, and ironically they became cruel oppressors of innocent minorities when they gained their chance, as Germans and Jews experienced when a free Poland was temporarily reconstituted after the First World War.

Marie completed her schooling with distinction at the age of sixteen. Meanwhile, her father had lost his fortune, so the daughters had to find ways to support themselves. Bronya and Marie had clear ambitions and strong ties to each other. They were very determined, very intelligent, and very Polish. Bronya went to Paris to study medicine, supported by her own savings and part of Marie's, while for about five years Marie worked as a governess—a position not much above a servant's—in several families, first in Warsaw, then in the countryside. Her meager savings went to Bronya with the understanding that as soon as Bronya finished her studies she would support Marie. So it happened that in 1891 Marie left Poland with a fourth-class ticket to Paris and forty roubles (twenty dollars) in her pocket. Imagine a Polish girl arriving in Paris in 1891 to study physics! It was a perfect romantic situation at a time when stories of Slavic heroines were quite fashionable. What made this particular heroine unique is that she had genius and truly meant to study physics.

Once in Paris, Marie enrolled in the Faculty of Sciences and attended courses in physics, chemistry, and mathematics with Professors Lippmann, Bouty, and Appell, the author of a famous treatise of "Mécanique Rationelle" [theoretical mechanics]. Outside the university she lived among Polish emigrés, one of whom was a talented young pianist named Paderewski. Marie's life was somewhat bohemian in its poverty, but her extreme discipline and severe working habits were anything but bohemian. She studied and worked fanatically, on almost no money and very little food. In 1893 she received the Alexandrowitch fellowship, a small grant of 600 roubles from a Polish organization. Typically, a few years later she saved her first earnings to pay back the fellowship. In 1894 a Polish physicist visiting in Paris, Joseph Kovalski, introduced her to Pierre Curie (1859–1906).

Pierre Curie (Figure 2.3) was the second son of physician Eugène Curie. The father, officially a Protestant, was a free-thinker with leftist tendencies. In 1871 he had participated in the Commune, not as a fighter but as a doctor, running a hospital in his home. Pierre seems to have been a strange

31

Figure 2.3 Maria Sklodowska Curie (1867–1934) and Pierre Curie (1859–1906), the most famous of scientific couples. [*Oeuvres de Pierre Curie* (Paris: Gauthier-Villars, 1908).]

child, considered slow mentally, but his father had enough sense to leave him alone to develop his talents freely and did not force him to go through the regular French schools. When he was about fourteen Pierre was placed under a tutor, who taught him mathematics and helped him with Latin. Pierre had a great facility for mathematics, and his intellectual development was then very rapid: At sixteen he was a bachelor of science. In 1883, at twenty-four, he was appointed chief of the laboratory at the School of Physics and Chemistry of Paris, a modest position he held for twenty-two years. He had a difficult character and was almost pathologically proud. The fact that he had not studied at the Ecole Polytechnique, from which most French scientists were recruited, stood in the way of a more brilliant career.

His early work was on the symmetry of crystals and piezoelectricity, which he discovered with his brother, Jacques. Then he extended the principle of symmetry and applied it to the study of many physical phenomena; in this respect his work on magnetism is still very interesting. Pierre Curie was the

first, to my knowledge, to introduce to physics ideas that today are called *group theory;* these included clear distinctions between polar and axial vectors, and the importance of symmetry in deciding which phenomena are possible. On the whole, Curie's early papers give the strange impression that he was a precursor of Eugene Wigner, and time has enhanced the importance of his point of view. By the time he met Marie Sklodowska early in 1894, he had already acquired a good scientific reputation. Lord Kelvin among others had recognized his talents.

A few months after his first encounter with the Polish student, Curie had made up his mind to marry her. On her part the decision was not without difficulties. Pierre was eight years older than Marie, not an unusual age difference at that time; however, she had never considered marriage, and the most serious obstacle was that, if she married, she would never return to live in Poland as she had planned. Yet when she went to spend her summer vacation in Poland, Pierre Curie's eloquent letters softened her. He was a

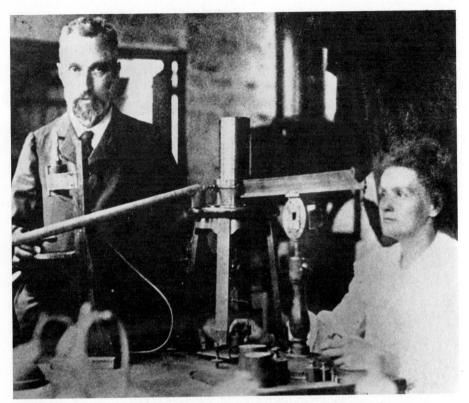

Figure 2.4 Pierre and Marie Curie in their laboratory with the quartz electrometer. (Courtesy of J. Hurwic.)

talented writer, almost a poet. They were married in July 1895, and spent their honeymoon pedaling through the French countryside on bicycles they had bought with money given to Marie for her trousseau. Throughout their life together their main relaxation consisted in exploring nature on bicycle trips.

At the time of their marriage Madame Curie had already passed what we would call her qualifying exam. Shortly after the birth of her first daughter, Irène, on September 12, 1897, she sought her husband's advice on a subject for a doctoral thesis. At his suggestion she undertook the study of "the new phenomenon" discovered by Becquerel. First she repeated Becquerel's experiments. But to measure the phenomenon with greater precision, instead of using Becquerel's electroscope, she used the apparatus shown in Figure 2.4, which had been devised by Pierre Curie. The feature worth noting in this apparatus is the compensating system, whose essential part is a piezoelectric quartz (remember that piezoelectricity had been thoroughly studied by Pierre and Jacques Curie). All electrometers used by the Curies in their subsequent experiments and all electrometers used in

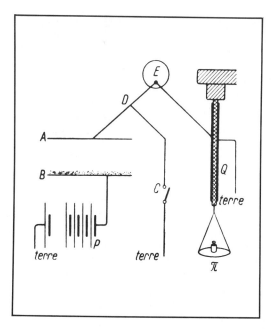

Figure 2.5 A schematic drawing of the electrometer, from the article published by Marie Curie in *Révue générale des Sciences* [*10*, 41 (1899)]. The instrument was used to measure the air ionization caused by radiations emitted by the radioactive substances placed in B. The condensor AB contained the radioactive substance on its lower plate. The electrometer E checked that the potential produced in the piezoelectric quartz Q by the weight balanced exactly the potential produced by the ionization current.

Madame Curie's laboratory after Pierre's death had this same feature. Otherwise, to her they were "not right." They actually were good instruments and worked well.

Using the electrometer (Figure 2.5), Marie Curie verified Becquerel's finding that the intensity of the uranium radiation was proportional to the amount of uranium in the compound and independent of its chemical form; it did not matter whether the sample was a uranyl salt, an oxide, or uranium metal. Marie Curie consequently confirmed Becquerel's finding that the emission of the rays is an atomic property of uranium. She then decided to examine "all the elements then known" and found that only thorium emitted rays "similar to those of uranium." At this point, after discovering that uranium was not the only element emitting spontaneous radiation, Madame Curie proposed the word *radioactivity* for this phenomenon. Also at this point she had a stroke of genius and decided not to limit her research to simple compounds of uranium and thorium but to also examine natural ores. She obtained many minerals from the collection at the Musée and set to work. As she expected, the minerals containing uranium and thorium were radioactive. To her surprise, however, the radioactivity of some of them was much greater—three or four times greater, by her method of measurement—than could be accounted for by their uranium or thorium content.

She drew the correct conclusion: Those ore samples must contain an element much more radioactive than uranium or thorium. Before announcing this hypothesis she made a test. She noticed that one particular uranium mineral, chalcolyte, was very radioactive when it was dug from the ground. She reproduced the chalcolyte with pure substances, out of bottles, and found that the artificial chalcolyte was no more active than any uranium salt. This proved that the natural ore contained a highly radioactive impurity. On April 12, 1898, her friend G. Lippmann presented her first communication to the Académie des Sciences, which was published in the *Comptes-rendus* under the name of M. Sklodowska-Curie. In it, she briefly described her experiment, advanced the hypothesis of the existence of a new radioactive element (not yet using the word *radioactive* in this paper), and related how she had tested this hypothesis.

If a new element actually existed, Madame Curie was going to find it. But how? She devised what she herself later considered the basic method of radiochemistry. Since she knew no chemical property of this element, only that it emitted rays spontaneously, she would have to start her search by looking for these rays. She would take a sample of ore, dissolve it if she could, separate it into its components by the standard methods of chemical analysis, and determine where the radioactivity went by making use of her electrometer. Madame Curie knew that the quantitative analyses that had been made on uranium ores at that time left uncertainties of about 1 percent by weight. If this 1 percent contained the unknown radioactive substance, and if the radioactivity of the ore was three times higher than the radioactivity attributable to its uranium content, the unknown substance had to be some three hundred times more radioactive than uranium. She underestimated this radioactivity: In fact, the substance with the strong radioactivity was not 1 percent of the sample, but perhaps one part per million, and its radioactivity per unit weight was not three hundred times but many million times that of uranium.

Undeterred by the immensity of the task she was undertaking, she set out to isolate the new element and began to grind sample after sample of ores. Very soon she recognized that she could not carry on alone and suggested to her husband that they join forces. Together, Pierre and Marie Curie began treating pitchblende and concentrating the most radioactive products, using her ingenious method of analysis.

Once they had concentrated the highly radioactive part in a very small residue, they decided that they had found a new element. It was far from pure, and in fact the element was only a small impurity out of their sample, but its radioactivity indicated the existence of a new element. They called it *polonium*—Marie Sklodowska-Curie couldn't have chosen any other name. In July 1898 the Curies presented a joint paper to the Académie des Sciences, announcing the discovery and explaining the method by which they had achieved it. The dates are important: In February 1896 Becquerel discovered radioactivity. At the end of 1897 Marie Curie became interested in "Becquerel's rays." By April 1898 she had determined what known elements were radioactive and she suspected the existence of new, more strongly radioactive elements. By July of the same year Pierre and Marie Curie had jointly discovered polonium. They also found that the substance disappeared spontaneously, reducing itself to one half in a characteristic time called *half-life*.

When Marie Curie had asked Pierre to join forces with her, he had been busy with his work on magnetism. He then shifted to radioactivity and remained in this field until his death in 1906. As for Madame Curie, all her work from 1898 to her death—36 years of work—followed the same pattern: more ores, higher levels of purification, greater concentrations. It was a single-minded task requiring unbelievable physical stamina, great intelligence, and the stubbornness that was characteristic of Madame Curie.

When the Curies had obtained their first highly radioactive product starting from pitchblende, they had carried out a chemical analysis of the mineral by the standard method. In the group of the sulfides, insoluble in acid solutions, they found polonium. But then they also found radioactivity in the barium group (barium, strontium, and calcium). At first they were unable to separate the radioactivity from barium; when they finally succeeded by fractional crystallization, they found a new, strongly radioactive substance, which they called *radium*. In September 1898 they announced the discovery of radium in their third paper, in collaboration with G. Bémont, a French chemist who aided them in the research leading to the identification of radium. Bémont had also helped the Curies find a laboratory; this laboratory was little more than a leaky shack, wet in winter and very hot in warm weather (Figures 2.6, 2.7). It had none of the features of a chemical laboratory now considered indispensable, such as hoods, and was completely unsuited for the task from a health point of view. However, at the time nobody knew of the hazards of radioactivity, and they had nothing better.

Professor Eduard Suess, a great Viennese geologist, graciously secured for the Curies a ton of residues from the pitchblende mines of Joachimstal, Czechoslovakia, which was then practically the only active center of uranium mining. The uranium had been extracted from the ores and what the Curies received were residues that nobody else cared for, but they knew that they contained radium. To those who know what it means to dissolve a hundred kilograms of an ore by hand, using crude methods, it will seem unbelievable that two people actually did it. Considering the fumes in a laboratory without hoods, the physical labor of stirring the solvents, and the heating of substances by primitive methods, only the indomitable spirit of the Polish woman could have overcome such odds. She undertook the most strenuous physical labor while Pierre, not as strong as she, attended to other tasks. Two years later they had enough radium in their sample to detect a change in the apparent atomic weight of the substance. Ordinary barium has an atomic weight of 137. The fractions richest in radium of their samples showed an atomic weight of 174. Madame Curie was striving for

Figure 2.6 The laboratory of the Ecole de Physique et de Chimie de la Ville de Paris where the Curies carried out the research that led to the discovery and isolation of radium. (From *Oeuvres de Pierre Curie,* Paris, 1908.)

pure radium, which, as we now know, has an atomic weight of 226. It took her twelve more years to obtain it, but already in 1900 she was on the right track and had made substantial progress at least in the methods if not in the final execution.

In 1900 Marie and Pierre Curie presented a report to an international physics conference held in Paris on "The new radioactive substances and the rays they emit." At the same conference Henri Becquerel also presented a report, dealing mainly with the radiations emitted. It seemed clear to him that, at least for some of the radiation, there was a strict analogy with x-rays and cathode rays. The Curies were primarily interested in the radioactive substances themselves. They described the methods for measuring radioactivity and the reasons why it should be considered an atomic and not a molecular property. They had studied the common radioactive sub-

Figure 2.7 Inside the Curies' laboratory. (From *Oeuvres de Pierre Curie,* Paris, 1908.)

stances, uranium and thorium; and later the new elements, polonium, radium, and actinium, the last of which had been discovered by André Debierne, a friend and associate of the Curies, in 1899. They then spoke of the chemical nature of the substances, their optical spectra, the effects of the radiations, and the so-called induced radioactivity (here, for the first and only time, they named Rutherford). Finally, they mentioned the great unsolved problems: the origin of the energy emitted and the nature of the radiations emitted.

We have seen that the new element, polonium, emitted particles and, of course, energy. It was even possible to measure the heat produced by it, and this presented a further difficulty for the physicists. By 1898 conservation of energy was well established, and scientists were quite unwilling to give it up. Where did the energy of polonium come from? All kinds of wild hypotheses were developed. One incorrect, yet not too unreasonable, idea was put forward by Marie Curie in her first paper on radioactivity (1898). It assumed the existence of rays going all over the world, in all directions, somewhat like neutrinos. For some mysterious reason these rays were not absorbed by anything except radioactive substances, which captured the rays and produced heat. Ernest Rutherford advanced the idea that radioactivity might depend on whether the sample was finely divided or not. The Curies themselves investigated whether heating or cooling a sample would increase or decrease its radioactivity—chemistry suggested such an experiment—but nothing happened. They also tried to compress their sample, but nothing they did would alter its radioactivity. It was a great mystery. In fact, the decay rate of radioactive elements was not altered until after World War II, and then by subtle chemical influences in which C. Wiegand and I used the understanding of the nucleus accumulated over the previous decade.

The Curie report was a small treatise on radioactivity as it was known in 1900; in 1904 Marie finally wrote her doctoral thesis, a similar but updated treatise.

By 1902 Marie had lost about 20 pounds as a consequence of her labor, but she had enough radium to enable her to measure its atomic weight. She found it to be 225, nearly the correct answer. She also had enough radium to observe its spectral lines. The report presented in 1900 contained some errors in the spectrum, but it correctly gave the two strongest lines, and there is no doubt that she had observed them. It was no wonder, however, that the couple started to suffer from strange and hard-to-diagnose diseases. Probably they had received enough radiation and had ingested enough radioactive substances to be seriously injured. Although Pierre died in an accident, Marie reached the age of 67, but she was ill for a long time and ultimately died of aplastic anemia, one of the conditions produced by overexposure to radiation. Her son-in-law, F. Joliot, examined Marie's experimental notebooks and found them strongly contaminated with radioactivity. Furthermore, Marie cooked at home and her cookbooks were still radioactive fifty years after she had used them.

Meanwhile, how had the Curies' careers progressed? One might expect them to have received some outside help after the years of tireless labor during which they had made memorable discoveries and lost their health. Major scientific figures such as Lord Kelvin had recognized their contributions, but the official French bureaucracy, responsible for their immediate livelihood, had not. Some friendly authorities arranged the Legion of Honor for Pierre, but he refused, saying he needed a laboratory, not a decoration. His friends urged him to accept the honor, pointing out that it would help him obtain a laboratory. However, Pierre remained proud and obstinate and received neither the honor nor the laboratory. In 1898 he applied for the chair of physics and chemistry at the Sorbonne but failed to get it. At that time his salary was modest—about 500 francs per month. He had two

Figure 2.8 Marie Curie with her daughter, Irène. The mother trained her daughter in scientific research and during the First World War took her along as an assistant in the radiological ambulance service she created. (Archives of the Institute of Radium.)

daughters to support, Irène (Figure 2.8), who became a scientist worthy of her parents; and Eve, who became a pianist and later was known for a very successful biography of her mother.

In 1900 the Curies had a stroke of luck. The University of Geneva offered Pierre a chair on very favorable conditions. He hesitated but eventually turned it down. However, this was noticed by the French authorities, who finally gave him a position at the Sorbonne in 1904. Pierre was very pleased, especially because he naively thought that a physics professor would have a laboratory. Alas, he was wrong. The chair did not provide the right to a laboratory, and at the time of his death Pierre Curie still had not been able to obtain an adequate laboratory. As the Curies continued working in deplorable conditions, the problem of a laboratory became an obsession. In 1902 Pierre proposed his candidacy for membership in the French Académie. A candidate was required to pay visits to members and carry out other formalities, which Curie despised. He reluctantly made the effort. However, the membership went to E. H. Amagat, known for the liquefaction of gases; in hindsight he cannot be compared with Curie. After this defeat

Curie no longer sought recognition in France. Nevertheless, in 1905, the year before his death, he was elected to the Académie.

In 1903 the Curies shared the Nobel Prize with Henri Becquerel. This greatly improved their financial state, but neither the Nobel Prize nor the 25,000 franc Osyris Prize they received later brought them the desired laboratory.

At the presentation of the Nobel Prize, Pierre Curie closed his lecture with the following words:

> On peut concevoir encore que dans des mains criminelles le radium puisse devenir très dangereux et ici on peut se demander si l'humanité à avantage a connaître les secrets de la nature, si elle est mûre pour en profiter ou si cette connaissance ne lui sera pas nuisible. L'éxemple des découvertes de Nobel est caractéristique, les explosifs puissants ont permis aux hommes de faire des travaux admirables. Ils sont aussi un moyen terrible de destruction entre les mains des grands criminels qui entraînent les peuples vers la guerre. Je suis de ceux qui pensent avec Nobel que l'humanité tirera plus de bien que de mal des découvertes nouvelles. [Les Prix Nobel en 1903]

> [It is conceivable that radium in criminal hands may become very dangerous, and here one may ask whether it is advantageous for man to uncover natural secrets, whether he is ready to profit from it or whether this knowledge will not be detrimental to him. The example of Nobel's discoveries is characteristic; explosives of great power have allowed men to do some admirable works. They are also a terrible mean of destruction in the hands of the great criminals who lead nations to war. I am among those who believe, with Nobel, that mankind will derive more good than evil from new discoveries.]

Curie's words are remarkable because they show an awareness then of problems we are debating today. They also demonstrate that the optimism of previous decades was beginning to fade.

At the time they received the Nobel Prize Pierre was 44 and Marie 36; both were in poor health and exhausted. They lived in great seclusion, enjoying the countryside as much as possible. Their social circle was composed mainly of old scientific friends: J. Perrin, P. Langevin, G. Urbain, A. Cotton, A. Debierne, G. Gouy, and others, all physicists or chemists. For a while they became interested in spiritualism, but they soon gave it up. They also knew some artists; at the Curies' one might have found both the sculptor Rodin and a well-known dancer from the Folies Bergère.

On April 19, 1906, a terrible tragedy befell Marie. While crossing a street in Paris, in the afternoon, Pierre was hit and killed by a carriage whose horse had bolted. Pierre's death deeply disturbed Marie, who suddenly found herself alone and overburdened with responsibilities and work and moreover with a great fame which she disliked. She withdrew, emphasizing her reserve and her tendency toward isolation. From 1900 to 1906, Marie had taught at the Normal Girl's School in Sèvres, and then had become Chef de Travaux at the Faculty of Sciences—that is, assistant to her husband. She was now appointed professor at the Sorbonne to replace her husband and, without a comment, Marie continued Pierre's course of lectures on radioactivity, picking up exactly where he had left off.

In 1911 she received a second Nobel Prize. Also about that time, Marie had an affair with P. Langevin, a private matter that was the subject of much scandalous commentary as Langevin was married and had children and his wife left him for a period. Langevin's son relates these events in a biography of his father.

I have mentioned Langevin (1872–1946) several times. He was a brilliant French physicist, who exerted a great influence in his time on French science. He had worked as a student under Pierre Curie and, in England, under J. J. Thomson. His main achievements were in theory, where his analysis of magnetic properties of materials is a classic of permanent value. During the First World War he devoted himself to practical problems of submarine detection and developed quartz electric oscillators, the ancestors of those used in modern watches. His great clarity of thought and power of expression made him a very influential teacher. A close friend of the Curies and of Einstein, he had extended international relations and in many ways became a semi-official representative of French science. He was also devoted to libertarian ideas and later in life became a militant communist. During the German occupation of France he participated in the resistance movements and had ultimately to flee his country. He returned after the war having lost relatives to the Nazi persecutions.

As a beginning student of physics I once met Langevin at the home of a distinguished mathematician in Rome, where he was the guest of honor. He gave me at once an eloquent comparison between the physics I was studying and what he had studied in his time. At his time thermodynamics, which appeared to him very abstract, was one of the main subjects. Were we not lucky that now physics could deal with such nice concrete models, doing away with so much abstract thought. Little did he suspect of what was in store in the next few years with the development of quantum mechanics. I was flattered by the attention such a famous man paid to an insignificant student, by his cordiality of approach, almost as a colleague. He certainly was most charming.

By the time of the First World War, Marie Curie was a world-famous celebrity, and the French government had finally allocated funds for the long-desired laboratory; however, it was not finished until after the war and was used only during a less active period of Marie Curie's scientific life. At the beginning of the war Marie Curie was indignant about the lack of radiological equipment in the field hospitals where it could be most useful, and she personally tried to remedy the situation by organizing an ambulance service equipped with x-ray equipment. She took along her eighteen-year-old daughter as an assistant. I can just imagine a French general arguing with Marie Curie.

Figure 2.9 Marie Curie conversing with R. A. Millikan at the Volta Congress held in Rome in 1931. Behind Millikan is R. Fowler; on the right, W. Heisenberg.

After the war an American journalist, Mrs. W. B. Meloney, managed to interview Marie Curie. This was a remarkable feat, because Madame Curie had shielded herself from journalists, autograph seekers, and generally any publicity. Mrs. Meloney apparently managed to elicit Marie Curie's sympathies. She promised that American women would provide a strong radium sample. Amazingly, although she had developed the technical processes used in the isolation of radium, Marie Curie had very little radium of her own or in her laboratory. In 1921 she made a triumphant trip to the United States, where President Harding presented her with the radium purchased with the funds contributed by American women.

At a very early date it was realized that radium could have an effect on biological tissue. One of the first to be burned was Henri Becquerel, who carried some radium prepared by the Curies in the pocket of his waistcoat.

The Curies also suffered from lesions caused by radium. The idea developed that this new substance could be useful in controlling tumors, and thus it was clear that radium would have commercial value. However, the Curies decided not to try to patent any of the procedures they had developed for isolating the substance. This conformed with their beliefs and has often been cited as a proof of a lofty outlook. In my opinion, although there can be different views on the ethics of patenting scientific discoveries, there is no particular virtue in refusing the benefits of one's own work.

Marie Curie later participated actively in international intellectual cooperation through the League of Nations. In 1929 she took another trip to the United States. Meanwhile, in the new laboratory, she pursued her life work on the old theme of disentangling the radioactive families. She was also concerned with the development of younger collaborators, including her daughter Irène.

However, Marie's health was failing. She underwent two operations for cataract that left her with poor vision. When I met her in 1931, she looked frail and pale (Figure 2.9). Her bandaged hands, damaged by radiation burns, twitched nervously. She had finally obtained the laboratory, but it had come too late. She also had a pension voted by the French national assembly and two houses, one in Paris and one on the Riviera. At the end, destiny was kind to her, and in the last months of her life, Marie Curie saw the great discovery of her daughter and son-in-law: artificial radioactivity. She died in 1934 in a sanatorium in the French Alps.

Chapter 3

Rutherford in the New World: The Transmutation of Elements

Although Ernest Rutherford (1871–1937) was a giant in physics, we must recall that he did not work in isolation and that his name also has symbolic significance. Progress in physics usually depends on many contributions by several scientists over a period of time until they ultimately come to fruition. The last investigator often gets most of the credit. Thus to avoid losing the thread in this book I must omit some important figures. It is not easy to follow all the interactions between research workers and to show what inspiration one physicist derived from another. For example, Becquerel, the Curies, and Rutherford read each other's results and they occasionally spoke to each other, but it is difficult to follow, and even more difficult to effectively describe, these very important exchanges. The citations in their respective scientific works give an idea of the intellectual exchange among the scientists.

Furthermore, some scientists have gathered "schools" around themselves. The relationship between teacher and disciple varies from case to case, but often the personalities complemented each other and made the work more productive. Other scientists are solitary; all the intermediate stages between these extremes exist. Rutherford was one of the most successful heads of a school, and his collaborators made many of the discoveries that came from the laboratories he directed. It is sufficient to mention O. Hahn and F. Soddy in Canada; H. Geiger, E. Marsden, N. Bohr, G. Hevesy, and H. G. Moseley in Manchester; and J. Chadwick, P. M. S. Blackett, J. D. Cockcroft, E. T. S. Walton, M. Oliphant, M. Goldhaber, C. E. Wynn-Williams, and others at Cambridge. It is easy to see how Rutherford's inspiration and leadership was enhanced and supplemented by the intercourse between his disciples.

Figure 3.1 Ernest Rutherford (1871–1937) at age twenty-one. [From A. S. Eve, *Rutherford* (London: Cambridge University Press, 1939).]

Rutherford's Early Career

Rutherford (Figure 3.1) was born on August 13, 1871, near Nelson on the South Island of New Zealand, the descendant of a Scottish family that had emigrated there. His mother was a school teacher and played the piano. To play the piano in the conditions prevailing in New Zealand in 1870 was a sign of both culture and perseverance. Rutherford's father was an ingenious and energetic man who changed trades several times. He was a farmer, then established a small factory, and eventually operated a prosperous flax mill. The family lived in a pioneering environment in a subtropical climate. Rutherford had six brothers and five sisters, three of whom died in childhood.

Young Rutherford went to primary school in New Zealand, and at age ten read Balfour Stewart's book on physics. This is not an uncommon beginning for physicists; as children they often find a physics book that fascinates them. In 1882 Rutherford's family moved to Pelorus Sound, and here Rutherford attended secondary school and then Nelson College. (When, nearly half a century later, Rutherford had to choose a title for the peerage, he selected Lord Rutherford of Nelson). He received from the college a scholarship of fifty-five guineas, enough to support himself for a year. He scored 580 out of a possible 600 points in his admission examination and was first in English, French, Latin, history, mathematics, physics, and chemistry. In 1889 he received another scholarship and went to Canterbury College, where there were seven professors and 150 students.

Figure 3.2 Rutherford's first laboratory. It was located in a basement of Canterbury College in New Zealand. Here Rutherford carried out research on the propagation of high-frequency alternating current in iron. He discovered that a magnetized steel wire can act as detector of an oscillating discharge, a fact that was known to others although Rutherford did not realize it. Thus he managed to transmit and receive radio signals over a short distance between an oscillator and a detector placed in a large shed, which was later demolished. This research, which Rutherford continued in Cambridge, led to the development of a magnetic wire hertzian wave detector. (University of Canterbury.)

It is a curious coincidence that both Rutherford and Marie Curie began their careers by studying the magnetization of iron. He did a bachelor of science thesis on the magnetization of iron by high-frequency discharges (Figure 3.2). It is remarkable that in 1889 a student in New Zealand could do such experiments for a thesis. This early work already shows Rutherford's characteristic touch. He wanted to demonstrate that the iron of a wire through which a high-frequency discharge had gone was magnetized only "skin deep." He magnetized the wire using the discharge and showed that it was magnetized. Then he dipped the wire in nitric acid, dissolving only the surface. The magnetization was gone. This investigation was later published in his collected works.

In 1894, when he was twenty-three years old, Rutherford competed for a scholarship funded by the 1851 London Exhibition. The scholarships had been instituted by the Prince Consort, the husband of Queen Victoria, who wanted to dedicate the funds remaining from the exhibition to scholarships for subjects of the British Empire. (These awards still exist and provide a great educational resource for the British Commonwealth. When, years later, it was proposed to abolish them, Rutherford exerted all his influence to preserve them). Rutherford came in second, but the winner declined and Rutherford got the scholarship, which allowed him to continue his studies in England. It is said that when the announcement of his prize arrived, Rutherford was on the family farm digging potatoes. He read the telegram bringing the news and said, "This is the last potato I have dug in my life." Rutherford had to borrow the fare for his trip. On the way Rutherford stopped in Adelaide to see W. H. Bragg (1862–1942), who also became one of the major physicists of the Empire. Bragg was slightly older than Rutherford and at that time was a young professor. Later, W. H. Bragg and his son, W. L. Bragg (1890–1971), made memorable discoveries in crystallography and in x-rays. W. L. Bragg succeeded Rutherford in both Manchester and Cambridge.

Rutherford arrived at Cambridge, England in September 1895 and was accepted as a research student by J. J. Thomson, who, as we have seen, had become Cavendish Professor at a very young age. At the time of Rutherford's arrival Cambridge University was undergoing important changes in its curriculum, giving more scope to experimental training and also opening its laboratories to foreign students. The first student to work under the reformed plan was Rutherford. His fame spread rapidly among his companions, who realized immediately that they were dealing with someone out of the ordinary. Not long after his arrival, a fellow student wrote, "We have got here a rabbit from the antipodes, and he's burrowing mighty deep."

There is a considerable collection of Rutherford's correspondence. He wrote to his mother regularly every two weeks for the rest of her life—and she lived to be ninety-two. While he was still in secondary school, he had become engaged to a schoolmate, Mary Newton, and he also wrote to her. Most of Rutherford's letters to his mother are lost, however, they may still exist; Mary saved those to her. Included in one letter is this description of Thomson as he appeared to Rutherford in 1895:

I went to the lab and saw Thomson, and had a good long talk with him. He's very pleasant in conversation, and he's not fossilized at all. As regards appearance, he's a medium-sized man, dark and quite youthful still—shaves very badly and wears his hair rather long. His face is rather long and thin, has a good head, and has a couple of vertical furrows just above his nose. We discussed matters in general and research work, and he seemed pleased with what I was going to do. He asked me out to lunch, to Scroop Terrace where I saw his wife, a tall, dark woman, rather sallow in complexion, but very talkative and affable; stayed an hour or so after dinner, and then went back to town again. ... I have forgotten to mention *the* great thing I saw: the only boy of the house, three and a half years old, a sturdy youngster of Saxon appearance, but the best little kid I've ever seen for looks and size. [Eve, *Rutherford,* p. 15]

The child was G. P. Thomson, the future discoverer of electron diffraction, who died in 1975.

At first, Rutherford pursued his studies in magnetism. In a letter he recounted how all was going well, how he had to report his studies in a seminar, and how everyone took him seriously. The work on magnetism soon found a practical application, as his discovery could be used in detection of wireless signals. Marconi was doing similar work in Italy, but there were fundamental differences between the two men. Marconi, more an inventor than a scientist, was interested mainly in the practical possibilities of the detector. Rutherford, although he had realized the potential applications of his studies, was not so interested in them. Nevertheless, the magnetic wire detector was of true importance and even caused legal questions over patents until the development of vacuum tubes made it obsolete.

Investigations in Radioactivity

Meanwhile, Röntgen had made his startling discovery, and everyone was rushing to work on x-rays. Rutherford, in collaboration with Thomson, began measuring the ionization produced by x-rays, and in 1897, after the discovery of radioactivity, he immediately applied his experience to the measurement of the ionization produced by uranium.

In a long work carried out at the Cavendish Laboratory in 1898 Rutherford realized that there were two kinds of radiation emitted from uranium; he called them *alpha* and *beta.* They are distinguished by their absorbability in matter. At that same time the Curies were discovering polonium and radium. They too, as well as Becquerel, were studying properties of the radiations emitted by radioactive substances. Their effort and that of Rutherford and of many other investigators overlaps in a complicated way. Mistakes were not rare, but in a few years the conclusion was reached that beta rays were cathode rays, that is, electrons, moreover, P. V. Villard in France found a more penetrating radiation, called *gamma,* which is similar to penetrating x-rays. However, alpha rays remained a mystery. Rutherford

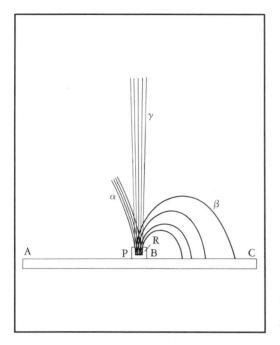

Figure 3.3 The three kinds of rays: α, β, and γ. They are distinguished by their trajectories in a magnetic field at right angles to the direction of motion. The α rays (helium nuclei) are positively charged; β rays (electrons) are much lighter and negatively charged; γ rays, analogous to x-rays, are quanta of electromagnetic radiation (photons). As they are neutral, they are not deflected by the magnetic field. Rutherford was responsible for the nomenclature. [From Marie Curie, *Thesis* (Paris: Gauthier-Villars, 1904).]

and the Curies both suspected that they were particles, atoms electrically charged and projected at high speed. Phenomenologically, they were characterized by their absorbability and by their small deviation in a magnetic field. Perhaps the reader is familiar with the famous figure frequently found in textbooks, which goes back to the Curies and Becquerel (Figure 3.3). It was also suspected that alpha rays had a definite range.

In 1898 a position became vacant at McGill University in Montreal, and Rutherford decided to apply. It was a momentous decision. To move from Cambridge, the world center of physics, to a colonial university in Canada was a leap in the dark. But Rutherford had an adventurous spirit; and, after all, he had come from New Zealand, which was not exactly the scientific center of the world. Moreover, Rutherford always, and justifiably, had great self-confidence. He obtained a letter of recommendation from J. J. Thomson, who wrote, "I have never had a student with more enthusiasm or ability for original research than Mr. Rutherford, and I am sure if elected, he would establish a distinguished school of physics at Montreal. I should consider any institution fortunate that secured the services of Mr. Rutherford as professor of physics" [Eve, p. 55]. Absolute truth; nevertheless, I have come across similar letters for young people who were not Rutherfords. Rutherford was appointed, went first to New Zealand to marry Mary Newton, and then left for Canada with her. His salary was 500 pounds a year, a comfortable sum and about double what Pierre Curie was earning. This reflects the conditions in the British Empire compared with those in the French Republic at that time.

Rutherford's new surroundings in Montreal were very congenial (Figure 3.4). He found newly built physics and chemistry laboratories supported and financed by a benefactor of the University, the millionaire Sir William Macdonald. The chairman of the department, John Cox, observed Rutherford for a few weeks and concluded "I think I'd better take your classes and do the teaching work. You keep on doing what you have to do." Here was a demonstration of outstanding intelligence and generosity. Rutherford became fond of his colleagues, and it is clear from his letters that he established most cordial relations with them, based on mutual appreciation of each other's contributions.

Rutherford obtained some radioactive substances and began to investigate the following problem. R. B. Owens and Rutherford had observed

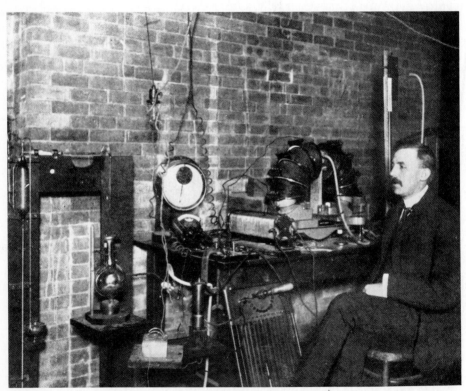

Figure 3.4 Rutherford at McGill University, Montreal, in 1905. In this year he began research on the properties of α particles. Their scattering later led him to the discovery of a heavy nucleus in the atom. (From Eve, *Rutherford.*)

that air currents influenced the ionization produced by radioactive substances. Rutherford suspected that they emitted radioactive gases. The Curies had already seen that objects placed near radioactive sources could acquire "induced radioactivity," which (we now know) is caused by active deposits given off from gases that escape the radioactive sources. If one does not take precautions, the formation of active deposits becomes a prime example of nonreproducible phenomena. Years later, I had such an experience at the time of the discovery of uranium fission, although I knew perfectly well about the possibility of radioactive gases and active deposits.

It was thus established that radioactive substances gave off radioactive gases as well as the three kinds of radiation. This discovery shook the foundations of chemistry, and one must read the original papers to appreciate the perplexity felt in those times. Rutherford said, "If they are gases, I'm going to show it beyond any doubt." Thus he began a very simple experiment, in true Rutherford style, using an open tube equipped with a series of electrodes, connected to an electrometer (Figure 3.5). He passed the gases, mixed with air, through the tube. The gas entered at one end and came out at the other. He measured the flow and thus the gas velocity. He also measured the radioactivity at the entrance and all along the tube, and its dependence on the velocity of the gas. From these measurements Rutherford could determine the half-life of radioactive gases. He also did other experiments to convince himself that the carriers of radioactivity were truly gases, because this conclusion seemed very strange.

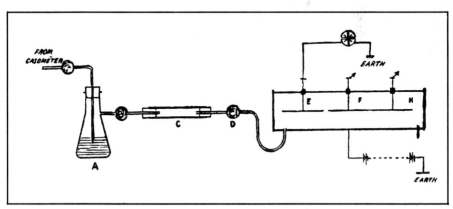

Figure 3.5 Apparatus for detecting thorium emanation and measuring its half-life. Air, passed through the material C, carries the radioactive gas to the tube on the right. The electrodes E, F, and H serve to measure the ionization caused by the gas. From this measurement and from the velocity of the gas, one obtains the half-life. [From the article by Rutherford and F. Soddy in *Transactions of the Chemical Society 81*, 321 (1902).]

At this point Sir William Macdonald gave the laboratory equipment for liquifying air. Rutherford passed air containing the radioactive gases through a copper tube cooled with liquid air to find out if the radioactive gases condensed. He found that they froze and that when the tube was warmed, they vaporized again. However, when he tried to measure the vaporization temperature, he made an error: He thought he had found a difference in behavior in the various "emanations," as he called the gases. I use the word *various* because he had gases originating from thorium or from uranium. Today we know the emanations are isotopic and therefore all have identical boiling and vaporization temperatures. But Rutherford found systematic differences in evaporation, which shows that even Rutherford could occasionally make a mistake. In any case, the discovery of the existence of the radioactive gases in itself was correct and very important. It also showed that somehow uranium and thorium atoms disintegrated and that among the fragments there were gases.

The next problem concerned the nature of alpha rays. Were they also a gas? Rutherford tested this hypothesis, but the result was negative. Then he tried another direction. If they were charged particles, what was their ratio of charge to mass, that is, their specific charge? Were they deflected in a magnetic field or not? Opinions on this important point varied. Rutherford disagreed, in print, with Becquerel, who at first thought they were not deflected by a magnetic field. Then both agreed that they were deflected. They considered various hypotheses and produced some mistaken experimental results, as one can see in Rutherford's *Collected Papers*. Finally, however, Rutherford reached the important conclusion that the specific charge for alpha rays was similar to that of ionized helium. He also found that when radioactive substances are heated, they emit helium, and that radioactive minerals do the same. All this work took time, but between 1903 and 1904 Rutherford became convinced that alpha particles were helium ions. It was now a matter of proving it.

Disciples and the Discovery of Transmutation

While the physical study of radioactivity was producing these surprising results, chemical studies were also revealing astonishing phenomena. Crookes discovered in 1900 that if iron hydroxide was precipitated in a uranyl salt solution, all the radioactivity went into the precipitate, and the uranium remained inactive. However, after a few days the precipitate lost its activity, and the uranium reacquired it. Similar phenomena occurred in other radioactive substances on different time scales—sometimes hours, sometimes minutes. It seemed as if there were sprites in the laboratories who were enjoying putting the day's carefully separated preparations back together again at night. It all seemed mysterious until Rutherford and Soddy, whom we will mention shortly, got the idea that radioactive substances transmuted into each other.

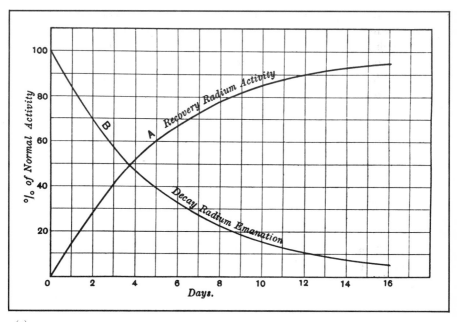

(a)

(b)

Figure 3.6 (a) Growth and decay curves of uranium X (^{238}Th). The curves show that the activity of UX extracted from uranium decreases, while the uranium, which had lost its activity at the time of extraction of UX, regains it, so the sum of the two activities is a constant. Later, these curves were used in Lord Rutherford's escutcheon (b). ((b) from Eve, *Rutherford.*)

Let us consider an example of transmutation. Uranium continuously produces by its decay another radioactive substance—UX—which in turn decays. In an undisturbed ore all this occurs at the same rate; that is, the same number of U and of UX atoms decay per second. However, if we chemically separate from uranium the active substance, UX, whose radioactivity is the only observable one due to the nature of the radiations, we find that the activity of the UX decreases with the half-life characteristic of UX because it is no longer replenished by uranium. At the same time the uranium, which had lost its radioactivity by the separation of UX, reacquires it. We thus have two activity curves—one for the separated substance and one for uranium. One falls and one rises (Figure 3.6). The sum of the activities is constant, because it corresponds to uranium in equilibrium or uranium as it is found in an undisturbed ore.

Rutherford and Soddy clarified all the complex phenomenology, reducing it to the fundamental rule that each radioactive atom has a definite probability (constant in time) of decaying in a unit time. This probability is characteristic of the substance considered and does not depend on anything else, as had already been clearly demonstrated by Rutherford in 1900. This was a brilliant and revolutionary idea, but it implied the transmutation of atoms, something which even Rutherford hesitated to mention because it sounded too much like alchemy. In fact, when he related to his Montreal colleagues what he had found and explained the phenomena, they advised him to be very prudent in presenting the evidence so that he did not make a fool of himself. Nevertheless, the facts were there and could not be denied.

We now come to Soddy. When Rutherford began to feel an urgent need of a chemist, one appeared in the person of Frederick Soddy (1877–1956). He had studied in England and completed his Ph.D. in chemisty. Looking for a position somewhere in the British Empire, he applied in Toronto, then wrote and wired, anxiously awaiting a reply. When no reply came, Soddy, a man of action, took a boat to Canada to find out what was happening. When he arrived in New York, he heard that the post he coveted had been filled. However, as he still had a ticket, he decided to visit Montreal. There he showed his letters of recommendation and asked for a post at McGill University. He was hired in the chemistry department. Rutherford, who needed help, spoke to Soddy, showed him what he was doing, and invited him to collaborate. Soddy promptly joined Rutherford. They soon became close collaborators, and from 1900 to 1903 they worked together establishing the theory of radioactive transmutations. Later Soddy contributed to establishing the concept of isotopism, according to which, in its primitive form, there may be substances chemically identical but different in their radioactive properties. Soddy also coined the term "isotopic" to indicate that the substances sit in the same place in the periodic system of the chemical elements. He became a professor of chemistry at Oxford, but in later years his creative genius waned.

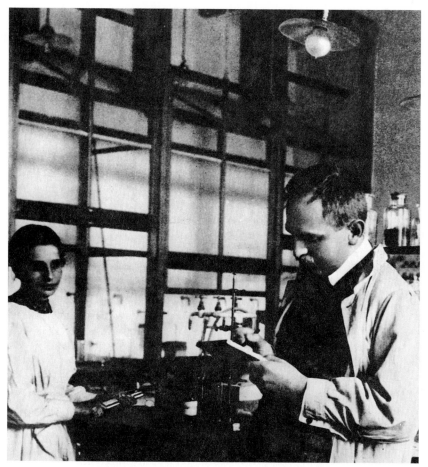

Figure 3.7 Otto Hahn (1879–1968). This distinguished German radiochemist, a student of Rutherford's, left his mark on all radiochemistry, from the discovery of new natural radioactive substances to the discovery of uranium fission. With him is Lise Meitner (1878–1968). (Ullstein Bilderdienst.)

After Soddy other students arrived in Montreal, including some future "stars," such as Otto Hahn (Figure 3.7). Hahn was born in Frankfurt into a family of businessmen. As a boy and as a student, he showed no signs of scientific precocity. He graduated in Marburg in 1901 after having studied organic chemistry and was accepted as an assistant to his professor, Theodor Zincke. Hahn had planned to have an industrial career in one of the German chemical firms that were then at their peak. Zincke suggested that he learn English well, and so he sent Hahn to work with Sir William Ramsay, the specialist in noble gases, who had discovered neon, krypton, and xenon.

Ramsay was trying to work in radiochemistry, without much success, and was not highly regarded by Rutherford. Nevertheless, Ramsay suggested an excellent problem for Hahn: Extract the radium from a certain preparation of barium chloride that contained radium. Neither Ramsay nor Hahn suspected that this preparation contained another isotope of Ra, Ra^{228}; the ordinary radium (the one discovered by the Curies) is Ra^{226}. Hahn eventually also found another radioactive isotope of thorium, which he called radiothorium (Th^{228}). R. B. Boltwood, a leading radiochemist at Yale University and a close friend of Rutherford's was not convinced by Hahn's work, and in a letter to Rutherford commented that radiothorium was "a compound of thorium and stupidity." Thus Hahn arrived in Montreal with a dubious reputation. To add to Rutherford's skepticism, Rutherford and Soddy had found another isotope of Ra, called ThX (Ra^{224}).

However, it did not take long for Hahn to convince Rutherford and Boltwood that he was right. Remember that at that time isotopism was unknown and thus each time a new radioactive substance was found, it was thought to be a new element, presenting a puzzle when it proved to be inseparable from an already known element. Hahn's radiothorium is an isotope (to use the modern term) of thorium and therefore is not chemically separable from thorium directly. However, since Th decays into MsTh, which is an isotope of Ra (Ra^{228}), and this in turn decays into RaTh, it is possible first to separate MsTh from Th and then, after RaTh has grown again in the MsTh, to separate the former into a relatively pure state. We thus obtain a separation of RaTh from Th, which would be impossible to carry out directly. We might marvel that Rutherford and Hahn did not grasp at the time the concept of isotopism, as they had discovered clear examples of isotopes; but when the mind is not prepared, the eye does not recognize, as we have seen in several cases.

Hahn remained with Rutherford for about a year. In 1906 he returned to Germany to Emil Fischer's Institute of Organic Chemistry, but he still continued to work on radiochemistry. In 1907 he joined forces with a young Viennese physicist, Lise Meitner (Figure 3.7), who was Max Planck's assistant, and they established an important center of nuclear research. After a few years Hahn obtained his own laboratory at the Kaiser Wilhelm Institute of Berlin Dahlem, still with Meitner. In 1938 Meitner had to flee from the Nazis, and she emigrated to Sweden a few months before Hahn and Strassmann discovered uranium fission. Hahn was then 59 years old, and perhaps this offers an extreme example of a discovery at a relatively old age.

Among the many marvels reported during this period, there was no shortage of false discoveries: "n" rays, with the tragic epilogue of the suicide of the man who believed he had discovered them; the transmutation of mercury into gold announced by attention-seeking chemists; and more. On the other hand, until 1904 the great D. I. Mendeleev, an old but always scientifically bold man, did not believe that helium could come from atomic transmutations.

Rutherford became much in demand as a lecturer and, as he enjoyed traveling, he went to many universities in the United States and the Commonwealth when his work permitted. Yale University offered him a position with a large salary, but Rutherford turned it down, commenting, "They act as if the university was made for students." Nevertheless, he maintained his close friendship with Boltwood at Yale. Their long and interesting correspondence, which has been published, gives a vivid picture of radiochemistry at the time, although it omits the important works of the Curies.

The origin of the energy liberated in radioactive disintegrations remained one of the great mysteries in the field. Rutherford, along with the Curies, A. Laborde, and others, measured this energy and found a surprisingly high number. He then thought that the radioactive materials in the earth might greatly influence its thermal balance, and this gave him the opportunity of settling an old argument between Lord Kelvin and the geologists. Lord Kelvin had calculated the cooling rate of the earth and from it inferred the time elapsed between the state of a molten incandescent sphere and the present. The result was much too short a period when compared to the geological evidence; hence, the dispute. When Rutherford considered the heat supplied by the radioactive substances in the earth, the difficulty was resolved. Rutherford was proud of this finding; he described the presentation of his findings thus:

> I came into the room, which was half dark, and presently spotted Lord Kelvin in the audience and realised that I was in for trouble at the last part of my speech dealing with the age of the earth, where my views conflicted with his. To my relief, Kelvin fell fast asleep, but as I came to the important point, I saw the old bird sit up, open an eye and cock a baleful glance at me! Then a sudden inspiration came, and I said Lord Kelvin had limited the age of the earth, provided no new source was discovered. That prophetic utterance refers to what we are now considering tonight, radium! Behold! the old boy beamed upon me. [Eve, p. 107]

In 1903 Rutherford was elected to the Royal Society of London, and on May 19, 1904, he delivered the Bakerian Lecture, titled "Succession of Changes in Radioactive Bodies." The Bakerian Lecture had been instituted by an amateur microscopist of the eighteenth century who left an endowment for it. The lecturer is called to review his own work, and delivering the Bakerian lecture is one of the highest honors that British Science may confer. Rutherford was Bakerian Lecturer a second time in 1920.

In the Bakerian Lecture of 1904 he listed the topics of his discoveries as follows:

1. Nomenclature.
2. Rate of decay of the excited activity of thorium and radium for different times of exposure to the emanation, and for different types of radiation.
3. Mathematical theory of successive changes.
4. Application of the theory to explain the changes in (a) thorium, (b) actinium, (c) radium.
5. Matter of slow rate of change produced by radium: comparison of the matter with the radio-tellurium of Marckwald.
6. Apparent radioactivity of ordinary matter, due in part to an active deposit of slow rate of change from the atmosphere.
7. Comparison of the successive changes in uranium, thorium, actinium, and radium.
8. Discussion of the significance of "rayless" changes in the radioelements.
9. Radiations from the active products. Significance of the appearance of β and γ rays in the last rapid change in the radioelements.
10. Difference between radioactive and chemical change.
11. Discussion of the experiments made to measure the charge carried off by the α rays.
12. Magnitude of the changes occurring in the radioelements.
13. Origin of the radioelements.

This is a summary of Rutherford's work until that time. He had good reasons when he strode around the laboratory loudly singing, "Onward, Christian soldiers!"

However, Rutherford's Canadian sojourn was coming to an end. His return to England coincided with a new phase of activity. We now must leave Rutherford for a while in order not to disturb the historical train of events.

Chapter 4

Planck, Unwilling Revolutionary: The Idea of Quantization

In the previous chapters I have dealt primarily with major experimental discoveries, such as x-rays and radioactivity. Our heroes were mainly physicists who had caught sight of these new worlds. Now we must turn to the other side of physics, which, though less accessible to the layman, is of equal importance. I refer to the development of theoretical ideas that were fundamentally new and as revolutionary in their own fields as the discovery of x-rays and radioactivity.

We see here the subtle play between theory and experiment that propels physics in its zig-zag progress between new facts and new theories. The ultimate goal of physics is to describe nature and predict phenomena. It is impossible to do this starting with a priori theories; we would be stymied after a few steps, and every error would be compounded and would send us further from the right path. On the other hand, using experiments alone, we would soon be lost in a bewildering array of disconnected facts without any hope of making sense of them. It is the combination of theory and experiment, brought about by the use of mathematics as a language, that permits the astounding results physics has attained. It is Galileo's immortal accomplishment to have clearly understood the power of this alliance and to have indicated ways of achieving it.

It often seems as if physics follows a predestined line and that great scientists have simply accelerated progress. If one scientist had not been there, another would soon have taken his place and found the same thing. One important exception to this is the discovery of the quantum of action, to which this chapter is devoted.

The Theoretical Pillars of Physics

At the end of the nineteenth century classical physics had reached admirable heights; its structure was harmonious and, to some extent, complete. Mechanics had been brought to maturity by Newton, and Lagrange had systematized it so that it seemed to offer a universal model. There was hope that every chapter of physics could be reduced to mechanics. However, this expectation was already fading, mainly because Maxwellian electromagnetism no longer seemed reducible to mechanics, as Maxwell himself had once hoped. In any case, the fabric of the universe seemed to rest on two pillars: mechanics and electromagnetism. Boltzmann had hinted at this idea when he inscribed his treatise on Maxwell's theory with Faust's question in Goethe's drama, "War es ein Gott der diese Zeichen schrieb?" [Was it a God who wrote these signs?] The biblical story of Genesis could be rewritten in modern language: "Let there be light" became

$$\nabla \cdot \mathbf{E} = 4\pi\rho \qquad \nabla \cdot \mathbf{B} = 0$$

$$\nabla \times \mathbf{E} = -\frac{1}{c}\frac{\partial \mathbf{B}}{\partial t} \qquad \nabla \times \mathbf{B} = \frac{1}{c}\frac{\partial \mathbf{E}}{\partial t} + \frac{4\pi\mathbf{j}}{c}$$

that is, Maxwell's equations that rule the electromagnetic field of light. And for the motion of celestial bodies,

$$\mathbf{F} = m\mathbf{a} = k\,\frac{mm'}{r^2}$$

dictated by Newton.

Such an extreme point of view is as naive as that of the fundamentalists who believe in a literal interpretation of the Bible. Furthermore, in spite of all the triumphs, there were some small "clouds," as Lord Kelvin intimated. It was difficult to reconcile Newtonian mechanics with Maxwellian electromagnetism, although both had been confirmed by innumerable experiments and Hertz had shown that Maxwell's equation covered the phenomena of light.

A third pillar of physics, perhaps the sturdiest, was the science of thermodynamics. This science is very different from mechanics and electromagnetism because it can be applied to all models, although it does not suggest any one in particular. This science in its classical form originated from a scientific study of the steam engine by Sadi Carnot (1796–1832). It was later completed by the discovery of the conservation of energy by Robert Mayer (1814–1878), Hermann von Helmholtz, W. Thomson, and systematized by R. Clausius. Classical thermodynamics is the consequence of two apparently innocuous propositions: (1) It is impossible to build an engine that creates energy indefinitely. This is called perpetual motion of the first kind, and the proposition affirms the conservation of energy. (2) It is impossible to build an engine that does nothing but take heat from a source at constant temperature and convert it into mechanical work. To use Lord Kelvin's words (1848), "It is impossible by means of inanimate material agency to derive mechanical effect from any portion of matter by cooling it below the temperature of the coldest of the surrounding objects" [W. Thomson, *Mathematical and Physical Papers*, vol. 1, p. 174 (Cambridge University Press, 1882)]. This would be perpetual motion of the second kind. Note that it would be practically as useful as a perpetual motion of the first kind because, for instance, it could solely take heat from the sea and transform it into work. If this were possible, all practical energy problems would be solved. From these two postulates one can derive quite unexpected and remote consequences. For example, thermodynamics predicts the lowering of the freezing point of water under pressure, which explains the flow of glaciers.

Thermodynamics has the same degree of certainty as its postulates. Reasoning in thermodynamics is often subtle, but it is absolutely solid and conclusive. We shall see how Planck and Einstein built on it with absolute trust and how they considered thermodynamics the only absolutely firm foundation on which to build a physical theory. Whenever they were confronted by formidable obstacles, they turned to it.

Thermodynamics distinguishes reversible phenomena from irreversible ones. An example of the former is the perfect elastic collision between two spheres as considered in mechanics; on the other hand, the expansion of a gas in a vacuum is irreversible. In our everyday experience, we have seen practically reversible phenomena proceed in both directions of time. For instance, in an elastic collision we could start from the final state, reverse all velocities, and come to the initial state. This is not possible for irreversible phenomena. Nobody has ever seen all molecules of a volume of gas spontaneously concentrate in a corner of a container. If we take a movie of a phenomenon, we can run it forward or backward in time. If the phenomenon is reversible, we do not find any criterion for telling whether the film is running correctly or backward, whereas if the phenomenon is irreversible, we see that there is one correct way of running the film. When a flame under a kettle of water cools the water until it freezes, we decide that we have observed an irreversible phenomenon backward.

Whether we can pass from one state of a system to another reversibly depends only on the initial and final states. The direction of time in irreversible phenomena is determined. The idea of reversibility may be refined and expressed quantitatively by the concept of entropy. The entropy of a system is a quantity that pertains to it in the same manner as its volume or energy, and can be measured by suitable experiments. An irreversible change increases the entropy of an isolated system, and thus entropy increases, or at least does not decrease, with time.

So far so good. But all mechanical and electromagnetic phenomena in the schemata of these two sciences are reversible. We are then faced with the following question: If all elementary phenomena are reversible and all physical phenomena are a combination of them, where does the obvious irreversibility of the macroscopic world come from?

This fundamental problem preoccupied some of the finest nineteenth-century minds for a long time. The new science of statistical mechanics that emerged from their effors introduced the concepts of probability, and thus bridged the gap between mechanics and electricity on one

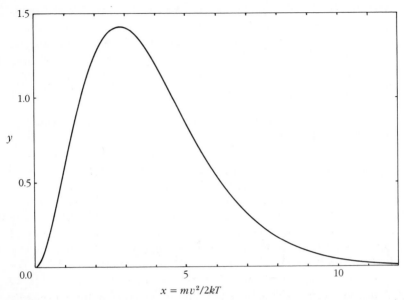

$$x = mv^2/2kT$$

Figure 4.1 (a) The graph of the velocity distribution of molecules in a gas, giving the number of molecules n in a unity velocity interval. This curve may be used for any molecular mass m and any absolute temperature T provided the scale of the abscissa is such that $x = (m/2kT)^{1/2}v$ and the scale of the ordinate is such that $y = (8m/\pi kT)^{1/2}n$. The equation of the curve is $y = x^2 \exp(-x^2)$. The most probable velocity is $(2kT/m)^{1/2}$ and the average kinetic energy of a molecule is $(3/2)kT$ independent of the mass m. These results were found by Maxwell.

side and thermodynamics on the other. Among the founders of statistical mechanics were Maxwell, R. Clausius, L. Boltzmann, and J. W. Gibbs.

It was found that the second law of thermodynamics has not an absolute validity, but an extremely high probability. This probability is so high that if we wanted to observe an exception on a macroscopic scale, we would have to wait a longer time than the age of the universe. For example, given many molecules in a vessel, nothing prevents them from accumulating in the left-hand half of the vessel. However, to see such an event we might have to wait for an extremely long time compared with 10^{10} years, the order of magnitude of the age of the universe. Thus the great Maxwell wrote to Lord Rayleigh, "*Moral.* The 2nd law of thermodynamics has the same degree of truth as the statement that if you throw a tumblerful of water into the sea, you cannot get the same tumblerful out again" [Rayleigh, *Life of Lord Rayleigh,* p. 47 (University of Wisconsin Press, 1968)].

The entropy considered by thermodynamics is a measure of the probability of the state of a system and also of the "disorder" of the system. Both probability and disorder tend to increase in an isolated system.

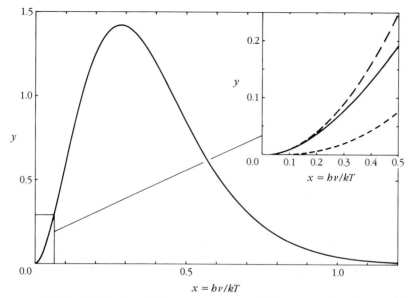

(b) For blackbody radiation at temperature T the energy density per unit frequency interval $u(\nu,T)$ is plotted. The curve can be used for any temperature, provided the abscissa has a scale such that $x = h\nu/kT$ and the ordinate has a scale such that $y = (\pi^2 h^2 c^3/k^3 T^3)u$. The equation of the curve is $y = x^3/(\exp x - 1)$. The limiting forms of Rayleigh and of Wien are shown in the inset, for small x.

Statistical mechanics allows us to tackle problems that go beyond the scope of thermodynamics. However, it requires more detailed models or other postulates and is therefore less secure and unchangeable than thermodynamics. Both thermodynamics and statistical mechanics are useful only when applied to problems involving many particles or, more technically, many degrees of freedom.

An important result of statistical mechanics was Maxwell's discovery of the law governing the velocity distribution in a monatomic gas. He found that the number of molecules in a given velocity interval is proportional to the density of the gas and depends only upon the temperature of the gas and the absolute mass of the molecules. We can calculate this velocity distribution with statistical mechanics, but not with thermodynamics. Maxwell discovered this law in 1859 by an intuitive but not rigorous reasoning. He gave better proofs later, but his original argument is strikingly clear and shows the power of his genius. A direct experimental confirmation was obtained only in 1921 by O. Stern. (See Figure 4.1.)

Statistical mechanics had much success and was studied intensely by Boltzmann and many other physicists after Maxwell. It gave rise to many subtle and difficult questions. A very remarkable theoretical result was that any degree of freedom of a system in thermal equilibrium has the average kinetic energy $\frac{1}{2}kT$. It was possible to prove this result rigorously on the basis of classical mechanics, and in many cases it was even verified experimentally, but in some cases it failed. Thus, for instance, the internal degrees of freedom of an atom did not manifest themselves in the specific heat, and a monatomic gas had a specific heat as if the atoms were points, which they certainly are not. These paradoxes troubled the students of statistical mechanics; Boltzmann, Lord Kelvin, Lord Rayleigh, and many others were puzzled by them. However, they did not suspect that the root of the trouble lay in nothing less than the validity of classical mechanics itself.

Possibly the most perfect treatise on classical statistical mechanics was written by Josiah Willard Gibbs in 1902, a slender book of considerable difficulty and subtlety titled *Elementary Principles of Statistical Mechanics*. The author, the greatest American physicist of the nineteenth century, possessed a deeply original and critical mind. He led a somewhat isolated life teaching at Yale University and was recognized by such people as Maxwell, but he was not particularly esteemed in the United States of his time. In the preface to his book he says,

> Moreover, we avoid the gravest difficulties when, giving up the attempt to frame hypotheses concerning the constitution of material bodies, we pursue statistical inquiries as a branch of rational mechanics. In the present state of science, it seems hardly possible to frame a dynamic theory of molecular action which shall embrace the phenomena of thermodynamics, of radiation, and of the electrical manifestations which accompany the union of atoms. Yet any theory is obviously inadequate which does not take account of all these phenomena. Even if we confine our attention to the phenomena distinctly thermodynamic, we do not escape difficulties in as simple a matter as the number of degrees of freedom of a diatomic gas. It is well known that while theory would assign to the gas six degrees of freedom per molecule, in our experiments on specific heat we cannot account for more than five. Certainly one is building on insecure foundation, who rests his work on hypotheses concerning the constitution of matter.
>
> Difficulties of this kind have deterred the author from attempting to explain the mysteries of nature, and have forced him to be contented with the more modest aim of deducing some of the more obvious propositions relating to the statistical branch of mechanics. Here, there can be no mistake in regard to the agreement of the hypotheses with the facts of nature, for nothing is assumed in that respect. The only error into which one can fall, is the want of agreement between the premises and the conclusions, and this, with care, one may hope, in the main, to avoid.

An Encompassing Problem: The Blackbody

There is another problem somewhat similar to the velocity distribution of a gas molecule—the blackbody problem. Classical thermodynamics can give some important limitations on its solution but cannot solve it completely.

Distinguished experimental physicists such as Heinrich Rubens (1865–1922), Ernst Pringsheim (1859–1917), Otto Lummer (1860–1925) of the Physikalisch Technische Reichsanstalt of Berlin (the German equivalent of the National Bureau of Standards) were working on it at the turn of the century, as were theoretical physicists elsewhere, such as W. Wien, Lord Rayleigh, and J. H. Jeans. Above all, a distinguished classical thermodynamicist, Max Planck, was wrestling stubbornly with it. I will first explain the problem and then speak of Planck.

A blackbody is a body that completely absorbs electromagnetic radiation falling on it. The opening of a cavity, such as the door of an oven, practically behaves as a blackbody. The emissive power, or emissivity, of a body is the electromagnetic power emitted per unit surface. The absorption power is the fraction of incident energy that is absorbed. For a blackbody the absorption power a is one because it absorbs all incident energy, by definition of a blackbody. The radiation is emitted in the form of electromagnetic waves of a certain frequency, and all frequencies are represented, each with its own intensity. We could examine the radiation with a spectroscope and find the energy in each frequency interval, obtaining the spectral distribution of the energy. The temperature of an oven can be judged by looking through a window into it. Already in 1792 T. Wedgewood, the porcelain maker and ancestor of Darwin, had noted that, when they are heated, all bodies become red at the same temperature. This crude observation was scientifically formulated in a precise way by Kirchhoff, who in 1859 proved by thermodynamics that the ratio of emissivity to absorption coefficient is a function of frequency and temperature only; it is independent of the nature of the body. For a blackbody that has an absorption coefficient of one, the emissive power depends only on the temperature and frequency, no matter how the blackbody is practically realized. Figure 4.2a shows a classical piece of apparatus used by the Reichsanstalt physicists for measuring the emissivity of a closed oven maintained at a certain temperature.

We can easily appreciate that finding the law governing the emissivity of a blackbody is an important problem. The fact that it is independent of the nature of the body points to a general character of the result, but even the scientists working on the subject could not suspect what fundamental and far-reaching consequences the investigation of blackbody emissivity was to have.

What could thermodynamics tell about the emissivity of the blackbody? The first and most important result is Kirchhoff's law mentioned above. Could it go further? Combined with electromagnetic theory, it could give more details. The total emissivity had to be proportional to the fourth power of the temperature. This had been found experimentally by the Austrian physicist J. Stefan (1835–1893) in 1879, and in 1884 Boltzmann derived this result by combining thermodynamics with Maxwell's theory of electricity. See Appendix 1.

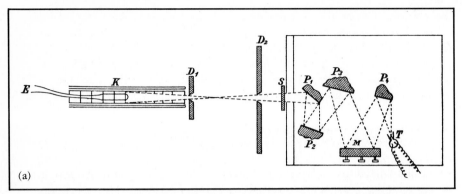

Figure 4.2 (a) Apparatus for measuring the intensity in the infrared of black radiation emitted from the oven K (blackbody). (b) Blackbody emission curves at constant frequency and variable temperature, compared with experimental data. [From H. Rubens and F. Kurlbaum, in *Annalen der Physik 4*, 649 (1901).]

A further step was taken by W. Wien in 1893. He could show, again by thermodynamics combined with Maxwell's theory, that the emissivity was the product of the cube of the frequency multiplied by a function of the ratio of frequency divided by temperature. From Wien's law it is easy to derive Stefan's law. Wien's law takes us as far as possible in combining thermodynamics with Maxwell's theory without introducing more detailed hypotheses. In fact, all attempts to find the exact formula for the emissivity, including important ones by Wien himself, failed. The results they gave were incompatible with experiment, or even absurd, because they predicted an infinite total emissivity. In place of the emissivity it is very often convenient to consider the energy density in the volume of the blackbody, for instance, inside an oven. The energy density can also be analyzed with respect to the frequency, and the energy density is simply proportional to the emissivity; the factor of proportionality is $4\pi/c$ where c is the velocity of light.

Max Planck

Let us now turn to Max Planck (Figure 4.3), who was to make decisive progress by finding the emissivity law and, through it, to open new unsuspected vistas on all of physics. Planck was born on April 18, 1858, at Kiel, Germany, a descendant of a family of lawyers and protestant ministers. His father was a distinguished professor of law. The Plancks were examples of the best qualities of the Germans: Honesty, devotion to duty, and perhaps also a certain stiffness of character were typical of the family. Planck's life

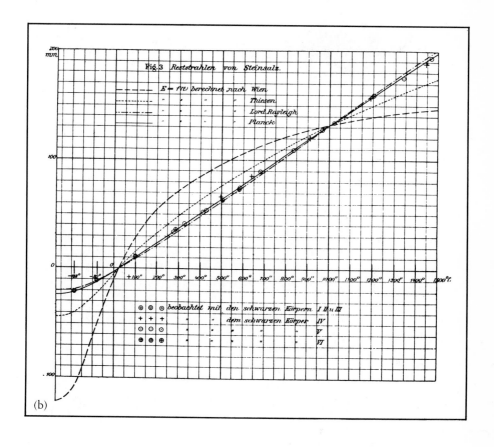

(b)

was marred by deep personal tragedies. His avocations were music, in which he showed professional accomplishment, and mountain climbing, which he pursued to a late age.

Planck's family moved to Munich, in Bavaria, in 1867, before the unification of Germany. Max attended the gymnasium at Munich and had a good physics professor, H. Mueller, who emphasized the principle of conservation of energy with striking examples that impressed the young pupil. At the University of Munich he attended rather insignificant courses and then, following the German custom of moving from one university to another, he went to Berlin, where he took courses from two celebrities: Kirchhoff and Helmholtz. The former delivered perfectly polished lectures that occasionally put the students to sleep. The latter did not prepare sufficiently and frequently could not be followed. Planck, like Heinrich Hertz, who had the same professors, bears witness to this in some letters to his family. Thus left to his own devices, Planck chose a dissertation on thermodynamics and reversibility and obtained his degree at Munich in 1879.

Figure 4.3 Max Planck (1858–1947) around 1900. His discovery of the quantum of action (1900) caused a revolution in natural philosophy, introducing discontinuity in many fields of physics and forcing a radical change in the description of phenomena.

The dissertation was important but was not noted. Planck said, "Eindruck Null" [No impression]. Helmholtz did not read it, Kirchhoff disapproved of it, and when Planck tried to submit it to Clausius, he could not be reached. In his thesis Planck applied some of his ideas to the study of thermodynamical equilibria and wrote a good fraction of what he later put in a famous thermodynamics book that became the standard text for several generations of physicists. However, Planck did not know that Gibbs had anticipated him in many ways. The American had published his results in the *Transactions of the Connecticut Academy of Sciences*—certainly a well-hidden place—and when Planck found out, he was disappointed.

Planck's first post was in his home town, Kiel. When Kirchhoff died in 1889, the University of Berlin called Boltzmann from Vienna as his successor. Boltzmann accepted but later changed his mind because he did not like the Prussian atmosphere of Berlin. After his refusal the University of Berlin rather unexpectedly called Planck. Helmholtz was the dominant figure at Berlin, and now Planck became acquainted with him as a colleague and developed great sympathy and a deeply respectful admiration for him.

In his autobiography he recalls that Helmholtz praised him on two occasions: for his theory of solutions and for his eulogy of Hertz.

Planck's love of fundamental and general problems drove him to take up the blackbody problem, which was independent of atomic models or other particular hypotheses. He loved the absolute, and such was the blackbody.

The discovery of the emissivity law was a goal worthy of his efforts, and thus Planck started to work on it in 1897. Wien, after having discovered some general properties of the emissivity law using thermodynamics, thought in 1893 that he had discovered it in its entirety and gave a formula for it. This formula seemed at first to agree with experimental data as shown in Figures 4.1c and 4.2b. It somewhat resembled the velocity distribution law for molecules in a gas and, in all, seemed credible. Planck tried for several years to derive it, rigorously combining electrodynamics with thermodynamics. However, Boltzmann showed that his reasoning was wrong because he did not do the correct statistical analysis required for calculating equilibrium conditions. Lively public arguments arose in which one of Planck's pupils attacked the mathematical foundations of statistical mechanics; Boltzmann answered pointedly, and he was right, as Planck himself had to acknowledge in the end.

In spite of these setbacks, Planck persevered in his work and reached an important result. Thermodynamics shows that the radiation in a blackbody does not depend on the nature of the walls but only on their temperature. Planck then thought to analyze a blackbody with walls made of hertzian oscillators whose behavior could be calculated. He thus avoided dealing with the detailed constitution of real molecules, which was unknown at that time. With this important simplification Planck found that the emissive power had to be proportional to the average energy of the oscillators. On the other hand, as Maxwell and Boltzmann had shown, statistical mechanics unequivocally required that the average energy of the oscillator be proportional to the absolute temperature. The formula:

$$\langle E \rangle = (R/A)T = kT$$

where $\langle E \rangle$ is the average value of the energy of the oscillator, R is a universal constant that appears in the relation between pressure, temperature, and volume of a mole of a perfect gas $pV = RT$, and A is Avogadro's number, that is, the number of molecules contained in a mole. In other words, for one molecule in a volume V, one would have $pV = kT$. Planck called the ratio $R/A = k$, Boltzmann's constant, and the name is universally accepted. Boltzmann, however, never used k in papers, writing it as R/A when needed. The combination of Planck's results on the relation between average energy of the oscillator and emissive power then gives

$$u(\nu,T) = \frac{8\pi\nu^2}{c^3} kT$$

as Rayleigh pointed out. This law did not seem to agree with experimental data and, furthermore, was obviously untenable because the total emissive power, integrated over all frequencies would be infinite.

Planck at the beginning of his work wanted to find a proof for a formula proposed by Wien: $u(\nu,T) = A\nu^3 e^{-\beta\nu/T}$, which seemed to agree with experimental data. As a master of thermodynamics, he postulated an expression for the entropy of the oscillators that would give Wien's formula and tried to show that his choice, which obeyed thermodynamics requirements, was unique.

Meanwhile, experiments started to cast some doubt on Wien's formula. The physicists at the Reichsanstalt were finding that, at low frequencies, for infrared radiation, the disagreement was greater than possible experimental errors. Planck was aware of these results and tried to alter his expression for the entropy of the radiation by generalizing it. See Appendix 2. From the new expression he calculated back the emissivity and found a formula, which he communicated on October 19, 1900 at the Physics Seminar of the University of Berlin. That same night, Rubens and Kurlbaum checked Planck's formula against their experimental results and found perfect agreement. Planck had found the blackbody formula. Perhaps it was only a fortunate interpolation, but it appeared to be correct.

Now it had to be justified theoretically. As Planck said in his Nobel Prize address twenty years later,

> But even if the radiation formula proved to be perfectly correct, it would after all have been only an interpolation formula found by lucky guess-work and thus would have left us rather unsatisfied. I therefore strived from the day of its discovery, to give it a real physical interpretation and this led me to consider the relations between entropy and probability according to Boltzmann's ideas. After some weeks of the most intense work of my life, light began to appear to me and unexpected views revealed themselves in the distance. [Les Prix Nobel en 1920]

What had Planck done? First, without difficulty, he had found the expression of the entropy of the radiation, corresponding to his formula for $u(\nu,T)$, but now he had to justify it. As long as he was dealing with classical thermodynamics, Planck was on familiar ground and a master, but the present task went beyond the scope of classical thermodynamics. He had to use the methods of statistical mechanics with which he was less familiar and, above all, in which he had much less confidence. But it was the only way to advance.

Between entropy and probability (setting aside some serious difficulties in the definition of probability) there is the relation first developed by Boltzmann (and engraved on his monument in Vienna):

$$S = k \log W$$

where W is the thermodynamical probability and k the universal constant that Planck called Boltzmann's constant.

The hertzian oscillators of the wall of the blackbody have a certain energy distribution and an entropy distribution. At equilibrium the entropy has to be maximum, and it can be statistically calculated using the fundamental Boltzmann equation. To calculate the probability by methods of combinatorial analysis, Planck found it convenient to divide the energy of an oscillator into small but finite quantities, so that the energy of the oscillators could be written as $E = P\varepsilon$ where P is an integral number. With this hypothesis Planck could calculate the average energy of an oscillator and thus find the blackbody formula. Planck expected that ε could become arbitrarily small and that the decomposition of E in finite amounts would only be a calculational device. However, for the results to agree with Wien's thermodynamic law, ε had to be finite and proportional to the frequency of the oscillator

$$\varepsilon = h\nu$$

where h is a new universal constant, appropriately called *Planck's constant*.

The energy density in the blackbody thus becomes

$$u(\nu,T) = \frac{8\pi\nu^3}{c^3}\frac{h}{e^{h\nu/kT} - 1}$$

where $h\nu$ is a finite amount, a quantum of energy. The harmonic oscillator could not have any energy as classical mechanics and electricity taught, but only discrete values integral multiples of $h\nu$. Planck's formula for $h\nu/kT \ll 1$ gives as an approximate expression the classical limit found by Rayleigh; for $h\nu/kT \gg 1$ it gives the formula found by Wien in 1893, and for intermediate cases it disagrees with both, but of course it is sustained by the experimental measurements (see Figures 4.1b (inset) and 4.2b).

The result is revolutionary. A competent and certainly not timid judge, Albert Einstein, said of it, "All my attempts to adapt the theoretical foundations of physics to these new notions failed completely. It was as if the ground had been pulled out from under one with no firm foundation to be seen anywhere, upon which one could have built" [Schilpp, *Albert Einstein, Philosopher-Scientist*, p. 45].

It was clear that the theory was neither a nightmare nor a pipe dream, because the experimental consequences were far reaching, correct, and concrete. Right from the first paper, Planck pointed out that from Stefan's law and from Wien's thermodynamical law it is possible to infer the two universal constants h and k, and from these the charge of the electron, Avogadro's number, and more. In Planck's paper of 1900 we find $h = 6.55 \times 10^{-27}$ erg · sec, and $k = 1.346 \times 10^{-16}$ erg/°C (Figures 4.4 and 4.5). Today we know that $h = 6.6262 \times 10^{-27}$ erg · sec and $k = 1.380 \times 10^{-16}$ erg/°C. The differences are minimal. From these figures Planck derived the charge of the

Figure 4.4 The title
page of the article by
Max Planck,
published in *Annalen
der Physik* [4, 553
(1901)], in which the
constant *h* appears for
the first time,
signaling the birth of
quantum physics.

9. Ueber das Gesetz der Energieverteilung im Normalspectrum; von Max Planck.

(In anderer Form mitgeteilt in der Deutschen Physikalischen Gesellschaft, Sitzung vom 19. October und vom 14. December 1900, Verhandlungen 2. p. 202 und p. 237. 1900.)

Einleitung.

Die neueren Spectralmessungen von O. Lummer und E. Pringsheim[1]) und noch auffälliger diejenigen von H. Rubens und F. Kurlbaum[2]), welche zugleich ein früher von H. Beckmann[3]) erhaltenes Resultat bestätigten, haben gezeigt, dass das zuerst von W. Wien aus molecularkinetischen Betrachtungen und später von mir aus der Theorie der elektromagnetischen Strahlung abgeleitete Gesetz der Energieverteilung im Normalspectrum keine allgemeine Gültigkeit besitzt.

Die Theorie bedarf also in jedem Falle einer Verbesserung, und ich will im Folgenden den Versuch machen, eine solche auf der Grundlage der von mir entwickelten Theorie der elektromagnetischen Strahlung durchzuführen. Dazu wird es vor allem nötig sein, in der Reihe der Schlussfolgerungen, welche zum Wien'schen Energieverteilungsgesetz führten, dasjenige Glied ausfindig zu machen, welches einer Abänderung fähig ist; sodann aber wird es sich darum handeln, dieses Glied aus der Reihe zu entfernen und einen geeigneten Ersatz dafür zu schaffen.

Dass die physikalischen Grundlagen der elektromagnetischen Strahlungstheorie, einschliesslich der Hypothese der „natürlichen Strahlung", auch einer geschärften Kritik gegenüber Stand halten, habe ich in meinem letzten Aufsatz[4]) über diesen

1) O. Lummer u. E. Pringsheim, Verhandl. der Deutsch. Physikal. Gesellsch. 2. p. 163. 1900.
2) H. Rubens und F. Kurlbaum, Sitzungsber. d. k. Akad. d. Wissensch. zu Berlin vom 25. October 1900, p. 929.
3) H. Beckmann, Inaug.-Dissertation, Tübingen 1898. Vgl. auch H. Rubens, Wied. Ann. 69. p. 582. 1899.
4) M. Planck, Ann. d. Phys. 1. p. 719. 1900.

Annalen der Physik. IV. Folge. 4.   36

electron and Avogadro's number. It was almost twenty years before these measurements were superseded. This is greatly to the credit of the physicists of both the Reichsanstalt and other Berlin institutions who measured the radiation constants and to the credit of Planck's theory, which established for the first time a link between such widely separated fields of physics.

Planck adds a poignant note: "It gave me particular satisfaction, in compensation for the many disappointments I had encountered, to learn from Ludwig Boltzmann of his interest and complete agreement in my new line of reasoning." However, although Boltzmann agreed, Planck's reason-

Hierbei sind h und k universelle Constante.

Durch Substitution in (9) erhält man:

$$\frac{1}{\vartheta} = \frac{k}{h\nu} \log\left(1 + \frac{h\nu}{U}\right),$$

(11)
$$U = \frac{h\nu}{e^{\frac{h\nu}{k\vartheta}} - 1}$$

und aus (8) folgt dann das gesuchte Energieverteilungsgesetz:

(12)
$$\mathfrak{u} = \frac{8\pi h\nu^3}{c^3} \cdot \frac{1}{e^{\frac{h\nu}{k\vartheta}} - 1}$$

oder auch, wenn man mit den in § 7 angegebenen Substitutionen statt der Schwingungszahl ν wieder die Wellenlänge λ einführt:

(13)
$$E = \frac{8\pi ch}{\lambda^5} \cdot \frac{1}{e^{\frac{ch}{k\lambda\vartheta}} - 1}.$$

Figure 4.5 Other excerpts from the same paper by Planck. Above, equation (12) contains the formula for the distribution of energy in the blackbody radiation as a function of frequency ν and temperature θ. The constant h appears in it as well as the velocity of light c and Boltzmann's constant k. Below, the numerical values of h and k in 1900 [formulae (15) and (16)]. Using these numbers, one can obtain the numerical values of the charge of the electron, of Avogadro's number, and of other universal constants in physics. These values stood unsurpassed for many years.

Hieraus und aus (14) ergeben sich die Werte der Naturconstanten:

(15)
$$h = 6{,}55 \cdot 10^{-27}\,\text{erg}\,.\,\text{sec}\,,$$

(16)
$$k = 1{,}346 \cdot 10^{-16}\,\frac{\text{erg}}{\text{grad}}\,.$$

Das sind dieselben Zahlen, welche ich in meiner früheren Mitteilung angegeben habe.

1) O. Lummer und E. Pringsheim, Verhandl. der Deutschen Physikal. Gesellsch. 2. p. 176. 1900.

(Eingegangen 7. Januar 1901.)

ing was subject to numerous serious objections. Such fundamental and revolutionary ideas were not easily assimilated. In spite of the various flaws mentioned above, the work was not ignored, but it was not at the center of attention. There were many spectacular discoveries at this time, and Planck himself was so diffident of the methods used that he spent years trying to explain his results in a less revolutionary way.

In 1931 the American physicist R. W. Wood asked Planck how he had invented something as incredible as the quantum theory. Planck answered, "It was an act of desperation. For six years I had struggled with the blackbody theory. I knew the problem was fundamental and I knew the answer. I had to find a theoretical explanation at any cost, except for the inviolability of the two laws of thermodynamics" [Armin Hermann, *The Genesis of Quantum Theory* (MIT Press, 1971), p. 23]. At the end of his life he commented further:

> My vain attempts to somehow reconcile the elementary quantum with classical theory continued for many years, and cost me great effort. Many of my colleagues saw almost a tragedy in this, but I saw it differently because the profound clarification of my thoughts I derived from this work had great value for me. Now I know for certain that the quantum of action has a much more fundamental significance than I originally suspected.

However, even at the beginning Planck was aware of the importance of his discovery. It is reported that, on a walk, Planck told his son that he had found something worthy of Newton.

As time passed, Planck became one of the most highly regarded German physicists. He was secretary of the Prussian Academy of Sciences and one of the most influential representatives of German science. Einstein, who did not sympathize with the German establishment, nevertheless had a deep respect for his colleague, even if they differed in political and scientific outlook. Their friendship was further reinforced by their common love for music, which they played together. Besides his scientific eminence, Planck's character inspired universal respect. A firm conservative, he found himself compelled by the strength of factual evidence and logical rigor to promote one of the greatest revolutions in natural philosophy.

Planck suffered grievous personal losses throughout his life. His first wife died in 1909, and of his four children by her, three died during the First World War. (The eldest son died at the front and his two married daughters succumbed in childbirth.) He later remarried and had another son. When he was seventy-five years old, Planck saw Hitler coming to power. For a German patriot such as Planck, not blinded by the dictator's propaganda and parades, this was a severe blow. At the request of his friends and colleagues, Planck accepted the presidency of the Kaiser Wilhelm Gesellschaft, now called the Max Planck Society, an important institution that supported a good fraction of German science. The burden was heavy and, under the circumstances, most unpleasant, but Planck thought it was his duty to try to save what he could. He even went to talk to Hitler in the hope of being able to moderate some of his worse aberrations, but he was shown the door by the Führer. Later the surviving son of his first marriage was put to death by the Nazis for having conspired against Hitler in 1944. Planck, by then very old, lost his house in an aerial bombardment and found himself caught between the retreating Germans and the advancing Allies. A German physicist heard of his plight and persuaded the Americans to send a car to take Planck to the relative safety of Göttingen. He survived the war, and Germany tried to indicate its emergence from barbarism by bestowing several honors on him. A great celebration had been prepared for his ninetieth birthday, but he died a few months before it, on October 4, 1947.

Chapter 5

Einstein—New Ways of Thinking: Space, Time, Relativity, and Quanta

In popular thinking Albert Einstein (1879–1955) is the incarnation of physics. In this case I think that popular opinion is correct and that, barring unforeseen developments in the remaining years of the century, he will be considered the greatest physicist of the twentieth century and one of the greatest of all time. If Raphael, returned to life, were to paint a modern *School of Athens* of physicists, I think he would include Einstein with Galileo, Newton, and Maxwell pointing toward the heavens, while Faraday and Rutherford would point toward the earth.

An Unconventional Youth

Einstein was born in Ulm on March 14, 1879, in a German Jewish family with liberal ideas. His father was an engineer but was not financially successful. Albert spent his childhood in Munich, and although he showed early signs of brilliance at home, he did not do exceptionally well at school. At secondary school he disliked the German teaching methods, and he came into conflict with the teachers, who in turn mistreated him. From these early experiences he developed a lifelong feeling of hostility toward official imperial Germany. Unfavorable business circumstances led the family to emigrate to Milan in 1894, and Einstein, who had been left in Munich to finish his studies, claimed to be ill and joined his family in Italy. He liked it better there, and during his short stay in Italy he took a walking trip from Milan toward Genoa, a distance of about 100 miles.

Einstein next applied for admission to the Polytechnical School in Zurich (Eidgenossische Technische Hochschule, or ETH) but was refused. Not only did he lack an adequate high school diploma but he also failed the admission examination, although he excelled in mathematics and physics. In order to gain entrance, he went to study at the gymnasium in Aarau. He was very happy there and fell in love with Switzerland; he later took Swiss

Figure 5.1 Albert Einstein (1879–1955). A picture dating from the period
when he was employed in the federal patent office in Berne and was writing the
immortal works that appeared in the *Annalen der Physik* in 1905.

citizenship, which he retained for the rest of his life. When he finally entered
the Polytechnical School in Zurich, his professors in mathematics were H.
Minkowski and A. Hurwitz, both first-class scholars, but he learned little
from them; nor did they notice him. However, Einstein had already begun
his solitary meditations on the major problems in modern physics, gathering
inspiration and information from his own readings. He became friends with
a fellow student, Marcel Grossmann, a Swiss who later became a university
professor of mathematics with a solid reputation. Einstein liked the practical
physics laboratory, where he could see phenomena with his own eyes rather
than through mathematical symbols.

When he graduated, he had difficulty finding employment to support
himself; at first he worked as a substitute teacher and gave private lessons in
physics. In 1902 the Grossmann family found him a modest job in the
federal patent office in the canton of Berne. About that time Einstein married Mileva Maric. They had two sons, one of whom became a highly respected professor of engineering at the University of California, Berkeley.

The job at the patent office was perfectly suited for Einstein (Figure
5.1). While he dealt with his duties in the office, examining the inventions

assigned to him, he also found time to think independently without being distracted. Later he advised young men to take similar jobs, as he felt they provided time for thought and so were appropriate for people with original ideas. He began writing articles on physics, which he sent to *Annalen der Physik,* then under the direction of W. Wien, of blackbody fame. He submitted one paper in 1901, two in 1902, and one each in 1903 and 1904. They were deep studies in thermodynamics and statistical mechanics. However, unknown to Einstein, the work had already been anticipated by Gibbs; the situation was similar to Planck's a few years earlier.

Then, in 1905, Einstein's genius flamed with unsurpassed brilliance. In March, May, and June he wrote three works, each of which alone would have sufficed to make him immortal. Only Newton, at age twenty-three, confined by the plague to his village of Woolsthorpe, had had a comparable spring. The first work—"Über einen die Erzeugung und Verwandlung des Lichtes betreffenden heuristischen Gesichtspunkt" [A heuristic point of view concerning the generation and transformation of light]—contains the discovery of light quanta and, as a minor application, the explanation of the photoelectric effect (Figure 5.2). The second—"Über die von der molekularkinetischen Theorie der Wärme geforderte Bewegung von in ruhenden Flüssigkeiten suspendierten Teilchen" [On the motions of particles suspended in liquids at rest required by the kinetic theory of heat]—contains the theory of Brownian motion, shows once again the real existence of atoms, and determines Boltzmann's constant in a new way. The third (Figure 5.3)—"Zur Elektrodynamik bewegter Körper" [On the electrodynamics of moving bodies]—contains the special theory of relativity, from which follows the famous formula $E = mc^2$, which is often all that the public knows of Einstein. In fact, there are some people who believe that this formula is the "secret of the atomic bomb"! It is a secret, however, that does not exist any more than the unicorn; both are figments of the imagination.

These works, on very different topics, show some common characteristics deriving from Einstein's scientific personality. They are revolutionary, open-minded, and they use simple mathematical methods. He reaches completely surprising results by applying uncompromising logic, firmly based on experiments.

Relativity

I will first discuss the third work, on relativity. From it came great practical consequences ranging from the balance of energy in an atomic bomb as well as in the sun, to the dynamics necessary for the construction of large particle accelerators. Above all, this paper caused a revolution in our concepts of space and time. For centuries philosophers had tried to analyze these concepts without ever achieving results as profound and definitive as Einstein's.

**6. Über einen
die Erzeugung und Verwandlung des Lichtes
betreffenden heuristischen Gesichtspunkt;
von A. Einstein.**

Zwischen den theoretischen Vorstellungen, welche sich die Physiker über die Gase und andere ponderable Körper gebildet haben, und der Maxwellschen Theorie der elektromagnetischen Prozesse im sogenannten leeren Raume besteht ein tiefgreifender formaler Unterschied. Während wir uns nämlich den Zustand eines Körpers durch die Lagen und Geschwindigkeiten einer zwar sehr großen, jedoch endlichen Anzahl von Atomen und Elektronen für vollkommen bestimmt ansehen, bedienen wir uns zur Bestimmung des elektromagnetischen Zustandes eines Raumes kontinuierlicher räumlicher Funktionen, so daß also eine endliche Anzahl von Größen nicht als genügend anzusehen ist zur vollständigen Festlegung des elektromagnetischen Zustandes eines Raumes. Nach der

**5. Über die von der molekularkinetischen Theorie
der Wärme geforderte Bewegung von in ruhenden
Flüssigkeiten suspendierten Teilchen;
von A. Einstein.**

In dieser Arbeit soll gezeigt werden, daß nach der molekularkinetischen Theorie der Wärme in Flüssigkeiten suspendierte Körper von mikroskopisch sichtbarer Größe infolge der Molekularbewegung der Wärme Bewegungen von solcher Größe ausführen müssen, daß diese Bewegungen leicht mit dem Mikroskop nachgewiesen werden können. Es ist möglich, daß die hier zu behandelnden Bewegungen mit der sogenannten „Brownschen Molekularbewegung" identisch sind; die mir erreichbaren Angaben über letztere sind jedoch so ungenau, daß ich mir hierüber kein Urteil bilden konnte.

Figure 5.2 Above, the title page of the first of the three great works of 1905, the paper formulating the hypothesis of light quanta. The manuscript had reached the *Annalen der Physik* on March 18, 1905. Below, the second great work: the paper formulating the molecular theory of Brownian motion, received by the journal on May 11, 1905.

To clarify this and other works of Einstein, I must briefly digress about light. The scientific study of light began in the Renaissance and reached a peak at the time of Newton and C. Huygens. These two major exponents of early post-Galilean physics had formulated conflicting theories on light. According to Newton, light was made of projectiles that moved very rapidly; according to Huygens, it was made of waves propagated in a very fine medium, the ether. These remained opposing theories for years, but at the beginning of the 1800s it was demonstrated that light could show interference phenomena; that is, by adding light to light, one could obtain darkness. This occurrence is easily explained by the wave theory, because

3. *Zur Elektrodynamik bewegter Körper;*
von A. Einstein.

Daß die Elektrodynamik **Maxwells** — wie dieselbe gegen-
wärtig aufgefaßt zu werden pflegt — in ihrer Anwendung auf
bewegte Körper zu Asymmetrien führt, welche den Phänomenen
nicht anzuhaften scheinen, ist bekannt. Man denke z. B. an
die elektrodynamische Wechselwirkung zwischen einem Magneten und einem Leiter. Das beobachtbare Phänomen hängt
hier nur ab von der Relativbewegung von Leiter und Magnet,
während nach der üblichen Auffassung die beiden Fälle, daß
der eine oder der andere dieser Körper der bewegte sei, streng
voneinander zu trennen sind. Bewegt sich nämlich der Magnet
und ruht der Leiter, so entsteht in der Umgebung des Magneten
ein elektrisches Feld von gewissem Energiewerte, welches an
den Orten, wo sich Teile des Leiters befinden, einen Strom
erzeugt. Ruht aber der Magnet und bewegt sich der Leiter,
so entsteht in der Umgebung des Magneten kein elektrisches
Feld, dagegen im Leiter eine elektromotorische Kraft, welcher
an sich keine Energie entspricht, die aber — Gleichheit der
Relativbewegung bei den beiden ins Auge gefaßten Fällen
vorausgesetzt — zu elektrischen Strömen von derselben Größe
und demselben Verlaufe Veranlassung gibt, wie im ersten Falle
die elektrischen Kräfte.

Beispiele ähnlicher Art, sowie die mißlungenen Versuche,
eine Bewegung der Erde relativ zum „Lichtmedium" zu kon-
statieren, führen zu der Vermutung, daß dem Begriffe der
absoluten Ruhe nicht nur in der Mechanik, sondern auch in
der Elektrodynamik keine Eigenschaften der Erscheinungen ent-
sprechen, sondern daß vielmehr für alle Koordinatensysteme,
für welche die mechanischen Gleichungen gelten, auch die
gleichen elektrodynamischen und optischen Gesetze gelten, wie
dies für die Größen erster Ordnung bereits erwiesen ist. Wir
wollen diese Vermutung (deren Inhalt im folgenden „Prinzip
der Relativität" genannt werden wird) zur Voraussetzung er-
heben und außerdem die mit ihm nur scheinbar unverträgliche

Figure 5.3 The third great work of 1905. Entitled "On the electrodynamics of moving bodies," it is the paper on relativity. It reached the *Annalen der Physik* on June 30, 1905, and was published in volume 17, p. 891. The two earlier works appeared in the same volume on pages 132 and 549 respectively.

two waves of opposite phases and of the same amplitude can destroy each other—the crests of one filling the troughs of the other. But interference is inexplicable with the corpuscular theory. Furthermore, the wave theory requires that in refraction phenomena light is propagated more slowly in denser media, whereas the opposite occurs in the corpuscular theory. In this respect, too, experiments favor the wave theory.

Later, Maxwell's equations, which seemed almost superhuman in their power and generality, accounted for all the known light phenomena. The propagation velocity could be related to electrical quantities measur-able in the laboratory by electrical instruments such as condensors, galvanometers, and magnets, which did not seem to have any connection with light. As light was reduced to electrical vibrations of the ether, the problem arose of determining the properties of ether. In some theories ether had been endowed with specific mechanical properties, such as a coefficient of elasticity. In any case, all the theories needed some sort of ether.

If a source of light or an observer moves with respect to ether, this movement should somehow manifest itself. In the case of sound propagated in air, when a source moves with the velocity of sound with respect to air at rest (mach 1), there are spectacular effects, which have been named the *sound barrier.* On the contrary, nothing could be observed that would show a motion relative to the ether. For instance, it had been found that the propagation velocity of light in a vacuum is always $c = 2.997924 \times 10^{10}$ cm/sec. This result had been precisely verified by A. A. Michelson and later by Michelson and E. W. Morley, who had carried out the measurements with one of Michelson's interferometers oriented in various ways with respect to the direction of the earth's movement.

Michelson (1852–1931) was born in Strelno (then in Prussia), but his family emigrated to California when he was a child, and he grew up in a gold mining town similar to those depicted in westerns. He went to the Naval Academy in Annapolis, where he graduated as a midshipman. His military training was followed by a long and brilliant scientific career that made him one of the most famous American physicists and the first American to win the Nobel Prize. His measurements of the speed of light are one of the pillars in the theory of relativity as it is usually taught, but they do not seem to have influenced Einstein in 1905.

Einstein was probably convinced a priori that Maxwell's equations had to have exactly the same form in all reference systems in uniform rectilinear motion with respect to each other. Perhaps he had reached his conclusions in his youth while meditating on how an electromagnetic wave would appear to an observer who was following it at a velocity c. In any case, for him the constancy of the velocity of light was indisputable. It had to be the same in all reference systems, although Einstein probably was not familiar with Michelson's direct confirmation, since he does not cite it in his paper of 1905. Einstein's axiomatic postulation of the principle of relativity recalls the principles of thermodynamics and states:

(1) *Postulate of relativity:* It is not possible to distinguish a reference system from another moving with a velocity constant in magnitude and direction with respect to it. Such systems are called *inertial.* The term to *distinguish* means that each experiment carried out in the first or second system gives the same result for an observer linked to the system.

(2) *Postulate of the constancy of the speed of light:* The speed of light is independent of the motion of its source.

Physicists before Einstein had struggled with the difficulties presented by the electrodynamics of moving bodies. Hertz, among others, had tried to give it a formulation. Lorentz, G. F. FitzGerald, and Poincaré had thought deeply about it. Lorentz had found that Maxwell's equations do not retain the same form when they pass from one reference system to another according to the rules given by Galileo and, apparently, based on common sense. Calling x the space coordinate and t, time in the first system; and x' and t' the corresponding quantities in the second system, which moves at a velocity v with respect to the first system, the equations of Galileo's transformation are:

$$x' = x - vt$$
$$t' = t$$

Transforming the coordinates from one system to another according to these rules, one would find observable changes in Maxwell's equations. Note in particular that the time is the same in both systems.

Lorentz had discovered a coordinate transformation, the famous Lorentz transformation, which leaves Maxwell's equations invariant and, when $v \ll c$, reduces to Galileo's transformation. It is expressed by the equations

$$x' = \frac{x - vt}{\sqrt{1 - (v/c)^2}}$$
$$t' = \frac{t - vx/c^2}{\sqrt{1 - (v/c)^2}}$$

Lorentz's transformation is correct, but at the time it appeared to be little more than a mathematical trick.

Einstein attacked the time and space transformation problem with a deep, almost childlike freshness of approach. With his powerful logic he carefully analyzed the concepts of time and space using an operational method that, for each concept introduced, required a rigorous and concrete specification of how to measure the magnitudes involved. Of course he was concerned not with the instruments and their mechanics, but with the logic of the experiment. Entirely unexpected results came from his analysis. For example, he discovered the relativity of simultaneity: Events occurring in different places and at the same time for one observer do not appear simultaneous to another observer who is moving with respect to the first. Similarly, there is the paradox of the twins: One twin remains in one reference system, and the other moves away in uniform and rectilinear motion, and then reverses the direction of movement and returns. When the second twin meets the first again, he finds him older than himself. If you are skeptical about this, the experiment has been carried out with particles that decay, and the result predicted by relativity was confirmed. Above all, it became clear that it was not Galileo's transformation but Lorentz's that corresponds to the correct way of measuring time and space.

Einstein's logic is irrefutable, although at the time it was distasteful to many physicists because of its apparent contradiction with everyday experience. I stress "apparent" because, in reality, the contradiction does not exist.

One of Einstein's professors, Hermann Minkowski (1864–1909), devised an elegant mathematical expression for the transformation of space and time coordinates. He introduced four-dimensional space, three dimensions of space and one of time. The special characteristic of this space is that the distances between two points is not given by Pythagoras's theorem, which, in ordinary space, says that $s^2 = x^2 + y^2 + z^2$. Minkowski's four-dimensional space uses a slightly different form: $s^2 = x^2 + y^2 + z^2 - c^2t^2$. It is the introduction of time in four-dimensional space and the minus sign in front of the last term that radically change the situation. In an epochal lecture in 1908 Minkowski introduced his new concepts with the following words:

> Gentlemen, the views of space and time which I wish to lay before you have sprung from the soil of experimental physics, and therein lies their strength. They are radical. Henceforth space by itself, and time by itself, are doomed to fade away into mere shadows, and only a kind of union of the two will preserve an independent reality. [Lecture to the 80. Versammlung Deutscher Naturforscher und Aerzte, Köln 1908]

When Minkowski saw Einstein's original work, he remembered his student and said, in effect, "Imagine that! I would never have expected such a smart thing from that fellow."

The consequences of the theory of relativity are vast, profound, and unexpected. In Einstein's way of thinking, the velocity of light appears as a universal constant, and its fundamental character transcends its historical connection to electromagnetism. It relates space to time in the Lorentz transformation; it is the limiting velocity that cannot be exceeded in the transmission of signals; and it appears in the connection between velocity, energy, and momentum of a material point having at rest the mass m. This connection is given by the formulae $E = mc^2/\sqrt{1 - v^2/c^2}$ and $p = mv/\sqrt{1 - v^2/c^2}$. The mass of bodies, defined as the ratio between force and acceleration, becomes variable with velocity. The conservation of mass is no longer a precise law and is substituted by a generalization of the conservation of energy, in which mass may transform into energy according to the formula

$$mc^2 = E$$

Modern physicists have even discovered massless "particles," the most notable being the light quantum (see next section). For these particles, $E = cp$, and they move with the speed of light with respect to any inertial system.

The new concepts disturbed physicists. Even such a great theoretical physicist as H. A. Lorentz, who had found the transformation basic to relativity, had difficulty in accepting the new ideas. It wasn't that the mathematics presented an obstacle; the difficulties in that field were minimal. The barrier was in the actual way of thinking, and consequently the theory only became familiar to physicists after a generation. Genuinely new ideas in physics penetrate slowly, mainly because the generation that creates them cannot "feel" them. Mature physicists can learn, but real assimilation occurs when the contemporaries die and the successors learn the new ideas as basics. I witnessed this phenomenon clearly in quantum mechanics. I should also add that the new generation, indoctrinated from the start, is unaware of many of the dilemmas and objections with which the creators had to wrestle.

Relativity was accepted only very slowly. For instance, even in 1922, the Swedish Academy awarded Einstein the Nobel Prize "for his service to theoretical physics and particularly for his discovery of the law of the photoelectric effect." It may seem strange that relativity is not mentioned, but, in retrospect, it seems to me that there was wisdom in this choice. It is not that relativity is a small matter but that Einstein's other "services" were immense.

Grains of Light and Molecular Hits

We now return to the first of Einstein's papers written in the amazing year of 1905. For me it is one of the greatest works in physics. At that time scientists knew that light was made of electromagnetic waves; if anything was certain, that was it. Yet Einstein doubted it and revealed the dual nature of light—corpuscular and wave. This discovery, along with the corresponding dual aspect of matter, became the greatest conquest of the century. Newton and Huygens were unexpectedly reconciled by a profound revolution in natural philosophy, which showed that both were partially right.

Planck had quantized the energy of the oscillators forming the walls of his version of the blackbody. With regard to the radiation proper, he could only give the expression $u(v,T)$. I believe he personally had no doubts that Maxwell's equations precisely described the electromagnetic waves that fill the cavity. Einstein, however, wondered if Maxwell's description was compatible with Planck's blackbody formula and reached the surprising conclusion that light itself must be composed of quasi-corpuscular quanta. I cannot leave out his reasoning because, when I read it, its power and simplicity struck me almost physically. (See Appendix 3.)

Einstein noted that Planck's derivation, in considering the energy exchanges between oscillators and radiations, "presupposes implicitly that energy can be absorbed and emitted by the individual resonator only in quanta of magnitude hv, i.e., that the energy of a mechanical structure capable of oscillations, as well as the energy of radiation, can be transferred only in such quanta, in contradiction to the laws of mechanics and electrodynamics."

Classical physics brings us unavoidably to the Rayleigh-Jeans approximation for Planck's formula. It is valid for $h\nu/kT \ll 1$, and in it h does not appear. For $h\nu/kT \gg 1$, Wien's formula is valid, and classical concepts fail. Einstein concentrated his attention on Wien's formula and from it derived an expression for the entropy of a radiation of a certain frequency contained in a certain volume, or, more precisely, for the variation of this entropy on changing the volume, keeping the energy constant. He notes that "this equation shows that the entropy of monochromatic radiation of sufficiently small density varies with volume like the entropy of an ideal gas. ..." He then proceeds to calculate the entropy by Boltzmann's method, concluding:

> If monochromatic radiation of frequency ν and energy E is enclosed in a volume v by perfectly reflecting walls, then the (relative) probability that at any moment all the radiation energy will be found in the partial volume v of the volume v_0 is given by
>
> $$W = \left(\frac{v}{v_0} \right)^{E/h\nu}$$
>
> From this we conclude further that monochromatic radiation of small energy density (within the validity range of the Wien radiation formula) behaves in thermodynamic theoretical relationships as though it consisted of distinct independent energy quanta of magnitude $h\nu$.*

Einstein takes this result very seriously, in spite of the overwhelming evidence from all propagation phenomena in favor of a wave theory of light, and says:

> If then as far as the dependence of entropy on volume goes, monochromatic radiation of sufficiently small density behaves like a discontinuous medium consisting of energy quanta of magnitude $h\nu$ it is reasonable to enquire if the laws of emission and transformation of light are so constituted as though the light were composed of these same energy quanta.
> We shall concern ourselves with this question in the next section. [Einstein, *Annalen der Physik 19,* 143, (1905)]

In the following section he elaborates on the photoelectric effect as well as on photochemical considerations and other ideas. All are examples confirming his hypothesis of light quanta.

The progress beyond Planck's ideas is enormous. Planck had only quantized material oscillators forming the walls of the blackbody, perhaps without even believing in the reality of energy levels. The whole tone of Planck's first, and even later, work gives the impression that quantization was little more to him than a calculational device. For Einstein, on the other hand, it was a fundamental phenomenon; in particular, light, that is, the

---

* For the convenience of the modern reader I have used the constants h and k. Einstein uses $\beta = h/k$ and $R/N = k$ with R the gas constant and N Avogadro's number.

electromagnetic field itself, is quantized. Naturally, quantization presents formidable difficulties if one tries to reconcile it with light propagation phenomena. Rather than flee from these problems, Einstein recognized their fundamental character and never stopped meditating on them until they were solved, partly by himself and partly by others.

We come now to the second of Einstein's papers of 1905. In the year 1827 the Scottish botanist Robert Brown (1773–1858) observed that pollen grains or other small objects suspended in water perform a random motion. This motion, called Brownian, is due to the impact of the molecules of the fluid surrounding the object. Einstein presents a theory of Brownian motion based on the kinetic theory of gases. It gives a new and direct method for determining Boltzmann's constant and thus Avogadro's number, and also an almost tangible proof of the existence of molecules. (See Appendix 4.)

From Patent Office to World Fame

These extraordinary works were noticed by the scientific community; in fact, as early as March 1906 Planck had published a paper on Einstein's relativity theory. However, no one had seen the author. Nor had the author left the patent office to go talk to major theoretical physicists. Nevertheless, a few adventurous young German scientists decided to go to Berne to find out who Mr. Albert Einstein was. They found him in his office. His home life seemed somewhat bohemian, but he was polite and ready to clarify any doubts about his ideas. Einstein began corresponding with Planck and Lorentz, and soon the Swiss authorities offered him a modest post at the University of Berne. At first they did not want to make him a *Privat Dozent* for formal reasons, but in 1908 the University of Zurich awarded him the title. The German University of Prague offered him a chair in 1909, and he accepted. However, Einstein was not happy in Prague, in the Habsburgs' Austria. He was disturbed by the sanctimonious and anti-Semitic atmosphere. Einstein had little sympathy for formal religion and was fundamentally agnostic, at least relative to all formal theology. He was relieved when in 1912 he was able to return to his beloved Switzerland, this time to the Zurich Polytechnical School where he had been a student.

In Prague Einstein had become friends with P. Ehrenfest; the friendship eventually became very close, lasting until Ehrenfest's death. Paul Ehrenfest (1880–1933) was an Austrian theoretical physicist who had been a student of Boltzmann. He had married a Russian physicist, Tatiana, and together they had written a famous article on statistical mechanics for the *Encyclopedia of Mathematics*. Ehrenfest had greater skill in clarifying obscure points in physics than in creating original theories. He was an exceptionally fine teacher, and showed great diligence in seeking to discover new talent.

He was known for the affection with which he encouraged young people, and he was loved by his many friends and students for his warm human qualities. He succeeded Lorentz in Leyden in 1912 and founded a flourishing school. Unfortunately, Ehrenfest was subject to periods of deep depression, and during one of them, in 1933, he committed suicide.

One physicochemistry student from a wealthy family joined Einstein in Prague. This was Otto Stern (1888–1969), whom I will discuss further later. Stern had understood that Einstein was the physicist of the future, or, rather, of the present, and he took advantage of his financial independence to study and work with Einstein. This is another example of the intertwining of the careers of many great physicists.

By now, Einstein had become an important professional physicist. In 1909 at a meeting in Salzburg, he met Planck, Wien, Sommerfeld, Rubens, Nernst, and other major modern physicists. He enjoyed being able to discuss problems face to face. Meanwhile, he had further developed his ideas of 1905.

As I mentioned above, by calculating the entropy of radiation and adopting Wien's law for $u(v,T)$, Einstein had reached the idea of light quanta and had linked it with experimental evidence. The entropy of radiation is strictly connected to the fluctuation of the energy contained in a given volume, and the light quanta hypothesis gives a very simple expression for such fluctuations. What happens if one takes up the calculation, using not Wien's approximate law, which is valid only if $hv \gg kT$, but Planck's exact one? Einstein did the calculation again and reached a remarkable formula for the fluctuation of energy contained in a fixed volume and in a certain interval of frequency around v.

Einstein found two terms that had to be summed. See Appendix 5. The first is the one obtained in 1905 with Wien's formula and is perfectly analogous to the expression of the fluctuation of the number of gas molecules in a volume. It indicates the granular structure of radiant energy and confirms that $E = nhv$, that is, that light behaves as if it were composed of quanta of energy $\varepsilon = hv$. The second expression on the other hand is exactly what one would obtain from pure electromagnetic theory and is caused by the constructive or destructive interference of waves. The remarkable fact that both are present indicates the dual nature of light—wave and corpuscular. The ideas of Newton and Huygens, which seemed mutually exclusive, are both confirmed. As Einstein wrote in 1909,

> It is undeniable that there is an extensive group of data concerning radiation which show that light has certain fundamental properties that can be understood much more readily from the standpoint of the Newtonian emission theory than from the standpoint of the wave theory. It is my opinion therefore that the next phase of the development of theoretical physics will bring us a theory of light that can be interpreted as a kind of fusion of the wave and emission theory. ...

Meanwhile, in 1907, Einstein had found another important application of quantum ideas. In 1819 P. L. Dulong and A. T. Petit had proclaimed that, according to their measurements, "Les atomes de tous les corps simples ont exactement la même capacité pour la chaleur" [The atoms of all the elements have exactly the same heat capacity], and Boltzmann had explained this fact with the principle of equipartition of energy. However, when it became possible to measure specific heats at low temperatures, thanks to the availability of liquid air and other cryogenic methods, Dulong and Petit's law was found to have many exceptions. Einstein gave the reason for the temperature dependence of specific heat by representing atoms in crystals as oscillators of a given frequency. (See Appendix 6.) According to quantum theory, they could only have the energy $nh\nu$ with n an integer. If $h\nu \ll kT$, we are in the domain of classical physics, and oscillators have average energy $3kT$. The atomic heat, that is, the specific heat referring to one atom, is $3k$, as Dulong and Petit had found. If, however, $kT \ll h\nu$, quantization becomes noticeable and we obtain a specific heat very different from the classical one. Qualitatively, when the oscillators are highly excited, the addition of a quantum of energy produces an energy change that is small compared with the energy already present in the oscillator, and we are not too far from the classical case, which assumes that the energy changes continuously. For such a situation to exist, the temperature of the oscillator's surrounding must be such that $kT \gg h\nu$. In fact, then the energy of the oscillator will be crudely kT, and this is large compared with the possible energy jumps $h\nu$. In the opposite limiting case $kT \ll h\nu$ the temperature agitation is insufficient to produce quantum jumps in the oscillator, which is thus unable to absorb energy from the surroundings. The behavior of the oscillator is dominated by its quantum properties, and the specific heat vanishes.

The theory was perfected by P. Debye and others and agrees with experimental data.

These considerations were important because they showed that the constant h plays a vital role in the mechanics of molecules and atoms. The success of these concepts served to bring even more attention to quantum ideas, which were still confined to a very restricted circle of initiates.

The World Order Collapses and Space Is Curved

In 1911 the first Solvay Council was held, on radiation and quanta (Figure 5.4). These conferences derive their name from Ernest Solvay, the inventor of an industrial method for the preparation of sodium carbonate, or soda. He instituted and financed a series of international physics meetings, with prearranged topics, to which the major physicists in the given field were invited. The conferences, limited to about thirty people, were held in Brussels. The format of the conferences and the list of invitees for the first one were devised largely by H. Walther Nernst (1864–1941), professor of physicochemistry at Berlin, one of the leading thermodynamicists of the time, and a power in German science. Nernst had discovered an important theorem, sometimes called the third principle of thermodynamics, according to which the entropy of any pure substance at the absolute zero was the same: zero. His theorem had deep roots in quantum theory. Because of his eminence in physics, linguistic and diplomatic abilities, and general prestige, H. A. Lorentz had a permanent invitation and often acted as chairman of the meeting. The restricted number of participants and the quality of those invited guaranteed lively and profitable discussions. By tradition, the Belgian sovereigns showed their interest in the meetings by inviting the participants to a dinner. As a result of these encounters Einstein established a friendship with Queen Elizabeth of Belgium and corresponded with her for many years.

At the 1911 conference, Planck, as always, expressed a conservative and prudent point of view, while Einstein, who was becoming recognized as the dominant figure, was more open minded. Shortly afterward, when Einstein was proposed for a chair at the Zurich Polytechnical School, Marie Curie and Poincaré, who had met Einstein at the Solvay meeting, clearly expressed their high opinion of him in a letter of recommendation.

Einstein went to the Polytechnical School in 1912, but his stay there was brief. The major physicists in Berlin wanted Einstein in the capital of the Reich, so they organized very attractive offers for him at either the Kaiser Wilhelm Institute or at the Prussian Academy, whichever he preferred. The teaching duties were minimal, and he was given maximum freedom for his work, a generous salary, and other advantages. Nernst and Planck went to Zurich to present the offer in person, a rare act in itself. Einstein took one day to decide. He informed the visitors that he would take a walk and would return with a rose—a red one if he accepted and a white one if he refused. He returned with a red rose. Einstein insisted on one condition: He wanted to keep his Swiss citizenship. This insistence led to some difficulties, for though Einstein considered himself Swiss, Prussia considered him Prussian.

Despite his brilliant position, Einstein was ill at ease in imperial Germany. He appreciated the company of his colleagues and other attractions of Berlin, but not Prussian militarism. To escape the barracks atmosphere he felt in Berlin, he often went to Holland to join his friends Lorentz and Ehrenfest (Figure 5.5). At the beginning of his German period Einstein was divorced; later he married a cousin, with whom he lived for the rest of his life.

We are approaching the fatal August of 1914. Einstein was busy beginning research on general relativity, a grand extension of the principle of relativity to arbitrary motions. It is clear that two systems, accelerated

Figure 5.4 The participants at the Solvay Council in 1911.
Standing, from left to right: O. Goldschmidt, M. Planck,
H. Rubens, A. Sommerfeld, T. Lindemann, M. de Broglie,
W. Knudsen, F. Hasenöhrl, H. Hostelet, T. Herzen, J. Jeans,
E. Rutherford, H. Kamerlingh Onnes, A. Einstein, P. Langevin.
Seated, from left to right: W. H. Nernst, M. Brillouin, E. Solvay,
H. A. Lorentz, O. Warburg, J. Perrin, W. Wien, M. Curie,
H. Poincaré. (Institut Solvay.)

with respect to each other, are not equivalent. In fact, inertial forces appear
in one which are not found in the other. A simple example is an elevator in
free fall on the earth's surface: To an observer in it, gravity vanishes. This is
obviously not equivalent to the same elevator at rest with respect to the
earth, for which gravity does exist. Einstein noted, however, that upon
introducing appropriate gravitational fields, one could render the two accelerated systems equivalent to each other. For this to be possible, it is necessary that the inertial masses that appear in the equation $F = ma$ be equal to

the gravitational mass that appears in the equation $F = kmm'/r^2$. This remarkable fact, often expressed by saying that the gravitational mass is equal to the inertial mass, had been discovered by Galileo in his semilegendary experiments in Pisa's leaning tower and confirmed by Newton, who carefully verified that pendulums of the same length, but of different materials have the same period. With much greater precision Baron R. von Eötvös confirmed this fact in Hungary, in 1891, and even greater precision was achieved by R. H. Dicke in 1963.

General relativity, as opposed to special relativity, was not at the forefront of physicists' interest, and Einstein confronted the problem alone. The general theory is considerably more difficult mathematically then the special theory of relativity, and it requires the use of "tensor analysis," then practically unknown to physicists, Einstein himself was impeded by the mathematical difficulties until his friend Grossmann introduced him to the works of B. Riemann and B. Christoffel, and, more important, to those of G. Ricci-Curbastro and T. Levi-Civita, which provided the necessary mathematical instruments. Einstein could then operate in curved spaces

Figure 5.5 From left to right: Zeeman, Einstein, and Ehrenfest in Amsterdam around 1920. Einstein and Ehrenfest were old friends, and it was probably during one of Einstein's visits to the Ehrenfests that he went to Zeeman's laboratory. (Museum Boerhaave, Leiden, Nederland.)

and tie gravitational effects to the curvature of space, which in turn is connected to the presence of matter or energy.

I am anticipating a little. The declaration of war in 1914 was accompanied by a general onset of naive and crude patriotism, especially in Germany. In response to the Allies' accusations, perhaps exaggerated at times, against Germany, especially concerning the invasion of neutral Belgium, German scientists answered with a defensive manifesto in which each sentence began with the phrase, "Es ist nicht wahr ... [It is not true ...], and at the end proclaimed the scientists' solidarity with the military. Unfortunately, among the manifesto's signers were highly respectable names such as Röntgen, Planck, Nernst, Wien, and many others. In the appropriate cases, they added titles to their names, such as *Geheimrat* [Privy councillor] or *Exzellenz* [His Excellency], an ironic detail if one thinks that someone like Röntgen could feel that such an official title, of all things, could lend force to his signature. But such was imperial Germany. It is clear that many signers, including Röntgen were naive, and they later regretted their action. Others, such as Planck, learned their lesson from the event. Though they remained solid German patriots, they were not swayed later by Hitler's nationalistic rhetoric. Einstein refused to sign the document and even considered organizing an opposing manifesto but then did not pursue the plan. In any case, he had already acquired serious enemies because of his political stand.

Einstein was absorbed in his effort to develop general relativity. Several times he thought he had achieved his goal and then discovered there were fatal flaws in this theory. Finally he succeeded in fulfilling his program of formulating physical laws in a way valid for any reference system, bringing about a new interpretation of gravity. The conceptual progress and the elegance of the theory are outstanding, though the experimental results are modest. It predicted a few small effects that could provide experimental evidence: the shifting of the perhelion of Mercury; the deflection of light rays by a mass, an effect visible in a solar eclipse; and the frequency shift of spectral lines emitted by massive stars. Unfortunately, all these effects are small and difficult to observe, so that even now there is no absolute and unequivocal experimental confirmation of the general theory of relativity. Nevertheless, the reduction of gravity to a geometric effect of the curvature of space is a magnificent concept and of immense importance in astrophysics and cosmology. It is here that general relativity comes into its own and becomes indispensable for describing and interpreting the astronomical observations.

In 1917 Einstein published another work of great importance, a new and very simple derivation of the blackbody law, which contained profound concepts. Light emission was no longer considered from a Maxwellian point of view. For the first time, statistical laws, the same as those applying to radioactive decay, were extended to electromagnetic phenomena.

Bohr's atomic theory already existed at that time, but the mechanism of light emission from an atom still remained a mystery. Einstein managed to lift a corner of the veil that shrouded the phenomenon, and his 1917 papers were very far sighted and signaled a milestone in modern theory. In them he abandoned the strict causality of classical physics and put at the forefront concepts of probability. Thus his work contained the germ of quantum mechanics and the profound revolution it caused. Moreover, without any explicit mention of the applications, it also contained the ideas basic to the workings of lasers and masers which we now use so much.

Here again, Einstein's profound idea is so simple that it can be stated in words, without formulae. For a more complete sketch, see Appendix 7.

Einstein considered a blackbody in thermal equilibrium containing in addition to the radiation especially simple atoms having only two energy levels. One can pass from one level to the other by emitting or absorbing a quantum of light of frequency $\nu = (E_1 - E_2)/h$ where E_1, E_2 are the energies of the two levels. By requiring that the atom and the radiation are in statistical equilibrium, that is, that the number of atoms going per second from the lower state to the upper state, or vice versa, is the same, he obtained remarkable relations between the transition probabilities and the emissivity of the radiation. He modeled the calculation on the chance occurrence of radioactive decay, suitably modified. This is a radically different approach from the classical electromagnetic method. Einstein called A the rate at which the atom spontaneously passes from the upper state to the lower, and

B the rate by which it passes from the upper to lower or vice versa under the influence of the radiation of emissivity $u(v,T)$. Among other things he showed that the two forced rates are equal. The *A* and *B* of Einstein are now a standard concept in physics.

The introduction of the *A* and *B* coefficients with their meaning based on probability, in connection with light emission, is an entirely new departure and a harbinger of things to come.

In the same paper of 1917, Einstein also gave a new proof of the result he had obtained years earlier, which said that in addition to the energy hv, light quanta also have a momentum hv/c in the direction of the propagation of light. Remarkably, he had reached this conclusion, which derives directly from relativity, by subtle and rather recondite considerations on fluctuations of energy and momentum density.

The Later Years and Einstein's Solitude

While Einstein was absorbed in these studies, the war had run its course, ending in 1918 with the defeat of Germany and the ruin of the Kaiser's old regime, with no regrets from Einstein. After a revolutionary period the new Weimar Republic brought great hopes for the democratization of Germany. In reality, however, the government was extremely weak, undermined inside and out, and incapable of governing with the necessary vigor.

In 1919 there was a total eclipse of the sun, which gave the possibility of verifying some of the consequences of general relativity. Various expeditions were undertaken for this purpose and the results, although not crystal clear, favored Einstein's theories. This was the point at which Einstein's popularity exploded. For reasons that are not clear to me, he suddenly became immensely popular, even with people who knew nothing of his works. He was treated like a movie star or a leading entertainer; he also quickly developed fierce enemies without any rational causes. There was even an anti-Einstein scientific society where once respected and respectable names became mixed with demagogues, madmen, and future Nazi recruits. Einstein was invited to give public lectures on relativity. He accepted, but the occasions were disrupted by his opponents and transformed into scurrilous political demonstrations. Einstein was not prudent and reacted in the press in a manner which, according to his closest friends, he should have avoided.

The situation took an ugly turn, especially since the extremists would not hesitate to assassinate their enemies. The murder of W. Rathenau, Minister of Foreign Affairs of the Weimar Republic, was a warning of what could happen. He was a great patriot and exponent of German industry who had worked indefatigably and successfully for the organization of the war economy and then had become a minister. He was a personal friend of Einstein,

Figure 5.6 Einstein making a point with the Italian mathematician F. Enriques in the courtyard of the University of Bologna, during a visit in October 1921.

and his assassination, along with those of the socialists K. Liebknecht and Rosa Luxembourg, and other notable figures, was only a prelude to the events of the Nazi era.

Einstein had had enough and left for a long trip around the world. His friends Planck, von Laue, and others urged him not to abandon Germany in such difficult times and not to accept the numerous offers he had from abroad. Leyden particularly attracted him both because of his friendship with Ehrenfest and because of Lorentz's urging that he go there. However, at the end of his trip, Einstein returned to Berlin after having established personal relationships in the United States and other countries. Things appeared to be calmer around 1924, and Einstein led a fairly active social life in Berlin. At his home he received a great variety of people, including the painter Sleevogt, the physician Plesch, the chemist F. Haber, the musicians F. Kreisler and A. Schnabel, the industrialist and foreign minister, Graf Rantzau, and the painter M. Liebermann (Figure 5.7). In his leisure time Einstein continued to play the violin, an activity he pursued throughout his life. He was not averse to playing the role of the great scientist; clearly, he enjoyed it. Perhaps this explains some of his affectations, his strange manner of dress, and some habits that may have been for show. After all, he was an admirer and friend of Charlie Chaplin.

However, none of these activities distracted Einstein from his very serious studies. The year 1922 had brought one more spectacular confirma-

Figure 5.7 Einstein as drawn by the renowned artist Max Liebermann (1847–1935) at Berlin in 1925. (By permission of the Institute for Advanced Study, Princeton, N.J.)

tion of his ideas on quanta when the American physicist Arthur Holly Compton (1892–1967; Figure 5.8) found that x-rays were diffused by free electrons as particles with an energy $h\nu$ and with a moment $h\nu/c$, as Einstein had predicted. In particular, the diffused quantum had a different frequency, which varied according to the angle of diffusion. These were things which could not be explained by wave theory.

The wave-corpuscle duality of light was becoming ever clearer when an unknown Indian physicist, Satiendranath (S. N.) Bose (1894–1974), submitted a manuscript to Einstein for his opinion. The manuscript contained a new proof of the blackbody formula, derived from statistical mechanics; however, Bose counted light quanta, to which he applied statistics, in a way different from the usual method. The essential difference is that, in Boltzmann's statistical mechanics, each molecule has individuality; thus we can assign each one a name and recognize it. In contrast, for Bose light quanta are absolutely identical. For Boltzmann trading the molecule A for molecule B gives a new configuration, which must be counted separately; but for Bose the exchange of two identical quanta does not give a new configuration. The result is a new proof of the blackbody formula. Einstein read the work Bose had sent him, translated it into German, and had it published in a German journal, adding a few words of praise and the comment that what Bose had done for light quanta, Einstein himself would demonstrate was possible also for molecules. Einstein kept his promise, and in 1924 he published a work on the subject. Thus were born the new statistics, which would be completed two years later by work of Fermi and

Figure 5.8 Arthur Holly Compton (1892–1967) in a photograph taken by F. Rasetti during an excursion on Lake Como organized for the participants in the Volta Congress in 1927. In 1922 Compton had done experimental studies on photon-electron collisions. (Courtesy of F. Rasetti.)

Dirac, who showed that Bose's and Einstein's statistics was not the only possible one and that in nature two types exist. We will discuss this question further in connection with Fermi.

As we might expect, Einstein also calculated the fluctuations of a Bose-Einstein gas and again found an indication of the dual nature not only of quanta, but also—and this is new—of molecules.

The wave-corpuscle duality was emerging more and more clearly from the shroud of mystery in which it was wrapped. Around that time L. de Broglie made decisive steps in the field. We will mention it later, but here too, Einstein's contribution was very important, especially in his role of validating new and surprising ideas.

As Einstein's fame grew, he made more frequent political and humanitarian pronouncements. He also took a clear stand as a pacifist and as a supporter of the State of Israel. When Israel was established many years later, Einstein was offered the presidency, but, knowing his own qualities and limitations, he declined. Any good cause could gain Einstein's support just as any good physicist he knew could obtain a recommendation from him. Thus, ironically, a recommendation from Einstein, which prima facie should have carried great weight, lost its value.

By 1927 nonrelativistic quantum mechanics could be considered complete, and the main lines of its interpretation, according to the so-called spirit of Copenhagen, had been formulated. Bohr gave a paper on it at the International Physics Conference at Como commemorating the Volta centenary. Einstein did not attend the conference because he did not want to visit fascist Italy, but a little later, when Bohr repeated the same ideas at a Solvay Council, he was opposed by Einstein. Bohr, who had been inspired by Einstein many times, worshipped him, and Einstein had given repeated expressions of admiration and affection for Bohr. Einstein's refutation of Bohr's dearest concepts hurt and troubled him. A long debate evolved, in which Bohr dismantled all Einstein's objections, only to be faced with new ones, without being able to convince him. To the end of his life, Einstein, one of the major creators of quantum physics, remained skeptical about the Copenhagen interpretation, although he became more and more isolated in this view.

With the arrival of Nazism, Einstein finally emigrated from Germany, where he would undoubtedly have been killed, and after some peregrination settled at the Institute for Advanced Study at Princeton, New Jersey. The flame of his genius was weakening, and Einstein, who for decades had seen farther ahead than anyone else and who had introduced some of the most profound and fruitful ideas in physics, devoted himself to problems that seemed to have no solution and perhaps were wrongly posed. The guiding light of new physics that had come from Berne, Zurich, and Berlin did not continue from Princeton.

Nevertheless, Einstein was still destined to play an important role, albeit a paradoxical one for a pacifist: urging the United States to construct the atomic bomb. I call it paradoxical, but his position conformed to his principles in the circumstances of the Second World War. But this was more a political than a technical step; Einstein certainly was not up to date in the nuclear physics of the 1940s, nor did he contribute technically to the development of atomic energy. On April 18, 1955, his life ended quietly in Princeton at age seventy-six.

He said of himself, "God is inexorable in offering His gifts. He only gave me the stubbornness of a mule. No! He also gave me a keen sense of smell."

Chapter 6

Sir Ernest and
Lord Rutherford of Nelson

Through the works of Planck and Einstein we have described the state of theoretical physics up to and beyond the First World War. But we left experimental physics at around 1907, when Ernest Rutherford left Canada and returned to England as a professor at the University of Manchester. From then until the First World War experimental research progressed rapidly, with Rutherford as the dominant figure. In this chapter we will deal with his work and other experimental discoveries of that period.

Back to England

The chair of Physics at Manchester, one of the main provincial English universities, had become vacant because Sir Arthur Schuster (1851–1934), a fine spectroscopist, had decided to retire from the position on the condition that Rutherford succeed him. Despite his German origins, Schuster was completely anglicized. He had inherited a fortune and had spent some of it equipping his department with an excellent laboratory. Although Rutherford's field of study was far from Schuster's, Sir Arthur felt he could not find a better successor, and he did his utmost to persuade Rutherford to accept. He had arranged an endowment for the institute and had also financed a fellowship for a theoretical physicist, which later benefited H. Bateman, G. C. Darwin (the naturalist's grandson), and Niels Bohr. Schuster also had a German assistant, Hans Geiger (1882–1945), who came under Rutherford's supervision, and later became a first-rate nuclear physicist. Geiger always remained in close contact with Rutherford and, after returning to his native Germany, he contributed to the blossoming of nuclear studies in his own country. His name is familiar to all who have dealt with radioactive substances, and have thus probably used his invention, the Geiger counter, one of the most useful instruments for detecting radioactivity. Also among the staff at Manchester was a technician, W. Kay, who helped Rutherford throughout his sojourn there.

Manchester, however, had little radium, less than 20 milligrams, and Rutherford needed radium, which, at the time was produced almost uniquely by the Joachimstal mines in Austria (now Czechoslovakia). The Vienna Academy of Sciences loaned 350 milligrams of radium bromide to Ramsay, at University College, London. It was meant for both Ramsay and Rutherford, but there was friction between the two scientists and they did not want to share the supply. Fortunately, Rutherford managed to obtain 350 milligrams more for himself as a loan from the Vienna Academy of Sciences, and thus the problem was solved. The radium from Vienna remained in Rutherford's hands during the First World War. At the end of the war the English government wanted to confiscate it as enemy property, but Rutherford insisted that the radium should be purchased and paid for as if it were a normal sale. With the proceeds he was able to help the impoverished Vienna Institute and its director Professor Stefan Meyer, who had been ruined by the war. This just and generous act profoundly moved Austrian scientists.

Having settled in Manchester and having obtained the necessary radium, Rutherford launched into research again. At first he reinvestigated old themes of research. Using spectroscopic methods, he obtained a definitive confirmation that the alpha particle was ionized helium. He did this with the help of T. Royds, another recipient of the 1851 exhibition scholarship (see Figure 6.1).

New Light on Alpha Particles

In 1908 Rutherford received the Nobel Prize in Chemistry. In his acceptance speech, "The Chemical Nature of the Alpha Particles from Radioactive Substances," he reported the counting of single alpha particles using the scintillation method. That is, he counted atoms one by one, looking through a low powered microscope at the flashes in zinc sulfide caused by the arrival of alpha particles. Geiger participated actively in this work, which was rather tedious and required them to spend long hours in absolute darkness.

If the reader wants to see scintillations, all he has to do is to look at the figures on a luminous watch dial with a magnifying glass. This must be done when the observer has been in the dark for some time, such as when he first wakes up in the dark. By counting atoms, Rutherford and Geiger had a means of determining Avogadro's number, the charge of the electron, and other universal constants that could also be found by entirely different experiments, for instance, by studying the blackbody radiation. The numbers derived from both methods corresponded very well, and these experiments convinced even the most skeptical physicists of the real existence of atoms, overthrowing the most obstinate, conservative rearguard.

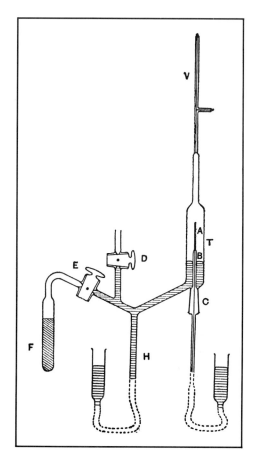

Figure 6.1 Rutherford and Royds' apparatus for demonstrating the nature of alpha particles. The needle A contains radon whose alpha particles emerge from the glass and fill the tube T with helium at low pressure. The helium, pushed by the mercury in the discharge tube V, shows the characteristic emission spectrum. [From *Philosophical Magazine 17*, 281 (1900).]

These experiments helped also to persuade the English of the merits of the quantum theory, as the electric charge of the electron and other constants determined by Rutherford were so close to the values given earlier by Planck, using blackbody radiation theory.

In counting, Rutherford and Geiger used quite an ingenious method that was later greatly expanded and advanced. The scintillations produced by alpha particles on zinc sulfide screen are not easily visible, and for each observer one can assign an efficiency defined as the ratio η between the number of scintillations that occur and the number observed. Assume that two observers (Rutherford and Geiger) watch the same screen, and each counts the scintillations he sees, indicating each by pressing a key like that used for sending a telegraphic message, thus marking also the instant of observation. There are three kinds of signals: (1) from the first observer alone, (2) from the second observer alone, (3) from both observers together.

Their number is:

$$n_1 = \eta_1 n \qquad n_2 = \eta_2 n \qquad n_{12} = \eta_1 \eta_2 n$$

where η_1, η_2 are the efficiencies of the observers; n is the unknown number of scintillations that occurred; and n_1, n_2, n_{12} are the numbers observed by 1 and 2 separately and together. It follows that $n = n_1 n_2 / n_{12}$.

Today, observing eyes are replaced by electronic devices or by Geiger counters, but the principle of the method is unchanged.

Rutherford faced another problem that seemed much more modest than counting atoms: describing and explaining the phenomena that accompanied the passage of alpha particles through matter. He attacked the problem with help of various students. Around 1904 W. H. Bragg with R. D. Kleeman had found that alpha particles with a given energy have a unique range, and the two Braggs (father and son) had studied the ionization along the trajectory. Ernest Marsden (1889–1970) a student from New Zealand who had come to work with his famous fellow countryman in 1909, by chance observed that occasionally alpha particles, instead of going straight or nearly straight, were deflected by matter and went at considerable angles. When Marsden related this observation to Rutherford, the Professor made him repeat the experiment to confirm it. The big deflections had greatly amazed Rutherford. He later said that it was as if someone had told him that having fired a pistol at a sheet of paper, the bullet had bounced back!

Several weeks passed. Then one day in 1911 Rutherford announced that now he knew why Marsden's particles were deflected at wide angles. And, moreover, he knew the structure of the atom.

The Atomic Nucleus

What had happened? At that time there were several atomic models. Lorentz used the idea that the electron was elastically bound to a fixed center, thereby explaining the Zeeman effect. There was Planck's oscillator and other models, including one by J. J. Thomson, who hypothesized an atom made of a positive electric charge diffused in a sphere with electrons interspersed, like raisins in pudding. However, such an atom, popular in England, could not scatter alpha particles at wide angles because if the alpha particle neared the center of the pudding, by penetrating it, it would be in a region of average electric field zero and thus could not be deflected. The same applied if it went far from the center, outside the atom.

Various scientists, including the Japanese physicist H. Nagaoka, had thought of the possibility of an atom built like the planetary system, but this was still vague and speculative. Rutherford provided a solid experimental basis for this theory, creating an atomic model that is still valid today. Although now the atom must be described in terms of quantum mechanics rather than classical physics, fortunately both give the same results in the case of Rutherford's experiments. Rutherford hypothesized that all the positive charge (Ze) and the mass were concentrated in a small volume in the center, which he called the *nucleus*. The nucleus was surrounded by Z electrons, which circled around it. The electrostatic attraction between the positively charged nucleus and the negatively charged electrons held the atom together. Rutherford said explicitly that he was not concerned with the stability of the system, a weak point, leading to grave difficulties. An alpha particle considered as a massive point charge, incident on the nucleus, is repelled according to Coulomb's law, and, as Newton had already calculated, it follows a hyperbolic orbit, with the nucleus as one of the focal points of the hyperbola (Figure 6.2). It seems that Rutherford had learned this as a student in New Zealand. The electrons, thousands of times lighter than the nucleus, do not affect the trajectory of the alpha particle. From this model one can determine the probability that the alpha particle will be deflected at a certain angle θ on crossing a material foil.

In concrete terms, the number of alpha particles that fall on a screen that subtends a solid angle $d\omega$ seen from the target, per particle incident on the target containing n atoms per unit volume and thickness t, is given by $nt\, d\sigma/d\omega$ with

$$\frac{d\sigma}{d\omega} = \left(\frac{2Ze^2}{mv^2}\right)^2 \cdot \frac{1}{\sin^4(\theta/2)}$$

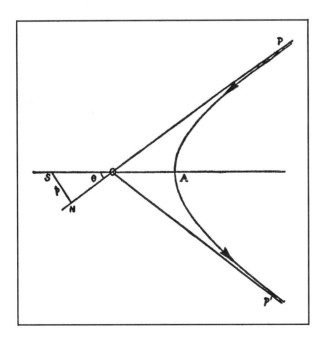

Figure 6.2 Trajectory (from P to P′) of an alpha particle deflected by a nucleus. Deflections of alpha particles passing through a thin metal sheet follow the law that was calculated on the basis of this figure. This proved the existence of a scattering charged center in the atom, later called the nucleus. [From Rutherford's article in *Philosophical Magazine 21*, 669 (1911).]

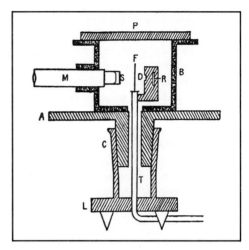

Figure 6.3 The apparatus used by H. Geiger and E. Marsden for studying the scattering of alpha particles. R is the alpha particle source, enclosed in a lead holder within the vacuum container B. A fine beam of alpha particles passes through the slit and strikes a thin metal strip, F. The alpha particles that pass through the strip strike the fluorescent screen S and are observed through the microscope M, which, with B, can be moved around TF. [From *Philosophical Magazine 25, 604* (1913).]

where θ is the angle of deflection, v the velocity of the particles, and m their mass. The mass of the nucleus is considered infinite with respect to m. Rutherford sent the manuscript describing his work to the *Philosophical Magazine* in April 1911.

Using the simple apparatus shown in Figure 6.3, Geiger and Marsden confirmed this formula in all its details by changing the substance on which alpha particles are projected, which changes Z, by varying the velocity of the alpha particles, by changing the thickness of the sheet, or by changing the angle of observation, θ.

The number Z, characteristic of the chemical nature of the target, was called the atomic number. It represents the electric charge of the nucleus, in units equal to the charge of the electron but with the opposite sign: For hydrogen $Z = 1$, for helium $Z = 2$, and so on. This discovery shed new light on the definition of a chemical element. To each element could now be associated a whole number, Z, that gives also the number of orbiting electrons. When, around 1869, Mendeleev with brilliant intuition developed his periodic system, he had ordered the elements according to their atomic weight. Now Rutherford's model revealed that it is the atomic number, not the weight, that counts. This was first noted by Antonius van den Broek (1870–926), a Dutch amateur scientist, in 1913.

The Planetary Atom

At this point it is quite difficult to follow in detail the nearly simultaneous discoveries of many features of the nuclear atom. We can get an idea of the state reached by reading, for example, a paper published by Rutherford in February 1914, called "The Structure of the Atom" (in *Philosophical Magazine* VI, *26*, 937). In it he discussed the scattering of alpha and beta particles and the conclusions that can be drawn concerning the localization of positive charges. He then gave special attention to the passage of alpha particles in hydrogen and to the collisions with what today we call protons or hydrogen nuclei. In a following discussion on the dimensions and constitution of the nucleus, he noted that Mr. Bohr had produced arguments for attributing nuclear origin to the electrons emitted in radioactive decay, thus distinguishing atomic electrons from nuclear ones. Regarding the nuclear charge, Rutherford related van den Broek's hypothesis as well as Soddy's law on radioactive displacement. The latter says that a nucleus changes its chemical nature according to its place in the periodic system, moving two steps back for the emission of an alpha particle and one step forward for the emission of a beta particle. This, of course, follows from the charge of the alpha particle $Z = 2$ and from the charge of the electron: -1. By emitting an alpha particle, the nucleus loses two units of charge; by emitting an electron, it gains one.

X-rays are useful in measuring the atomic number using either methods based on diffusion first suggested by W. Barkla, or the very powerful methods just invented by H. G. J. Moseley. Finally, there is the following passage:

> Bohr has drawn attention to the difficulties of constructing atoms on the "nucleus" theory, and has shown that the stable position of the external electrons cannot be deduced from the classical mechanics. By introduction of a conception connected with Planck's quantum, he has shown that on certain assumptions it is possible to construct simple atoms and molecules out of positive and negative nuclei, e.g. the hydrogen atom and molecule and the helium atom, which behave in many respects like the actual atoms or molecules. While there may be much difference of opinion as to the validity and of the underlying physical meaning of the assumptions made by Bohr, there can be no doubt that the theories of Bohr are of great interest and importance to all physicists as the first definite attempt to construct simple atoms and molecules and to explain their spectra.

For the sake of clarity I will refrain from giving more details on Bohr's work until the next chapter.

Same But Different: The Concept of Isotopism

At about this time it was becoming increasingly clear that atoms existed that were identical chemically but different in terms of radioactivity. Already in 1906 Boltwood had demonstrated at Yale that ionium could not be separated from thorium. In 1910, W. Marckwald, F. Soddy, and O. Hahn had shown the same inseparability for MsTh (discovered by Hahn) and radium, and further similar examples were found. In 1912 Rutherford gave the problem of separating RaD from lead to two of his research workers, G. de Hevesy and F. A. Paneth, Austro-Hungarians who were visiting Rutherford's laboratory. He told them, "If you are chemists worthy of your salt, separate them." After two years of work in which they tried everything, the two unfortunate men gave up.

However, de Hevesy and Paneth transformed their defeat into victory by inventing the tracer technique, which, enriched by the later discovery of artificial radioactivity, became one of the most powerful techniques in modern science, perhaps comparable in importance to the miscroscope. This is how it works: Given substances that are identical chemically but distinguishable by their radioactivity, one can carry out experiments in which the radioactive atoms can always be recognized even after complicated reactions have occurred. For example, if one eats ordinary salt mixed with salt containing radioactive sodium, an examination of the radioactivity in a drop of blood or in a urine sample will reveal how much of the ingested sodium was found in the measured sample. The ability to recognize atoms labeled by their radioactivity allows the solution of very important problems that would otherwise be totally inaccessible.

Thus the concept of isotopism in radioactive substances was being established; Soddy coined the name in 1913. It indicates "same place" in the periodic system. Isotopism was soon to be extended to stable nuclei. In 1912 J. J. Thomson found indications that the phenomenon of isotopism was not limited to radioactive elements, but was a general property shared by many. He measured the ratio of charge to mass for positive ions with the so-called parabola method. In this method perpendicular deflections of an ion beam are produced by means of a magnetic field and an electric field that are parallel to each other and perpendicular to the velocity of ions with a charge e and mass m. If the electric field E and the magnetic field B are directed along the x axis and the ions are moving in the z direction, the deflections are such that ions with the same ratio e/m, independent of their velocity, fall on a parabola located in a plane perpendicular to the beam velocity. (See Appendix 8.)

When J. J. Thomson applied this method to neon, he found that the element contained ions of masses 20 and 22 times those of hydrogen. Thomson considered the possibility that ions of a mass 22 could be caused by the compound NeH_2, but he found that this hypothesis presented great difficulties. In 1913 various attempts at fractionating Ne into different isotopes were only partially successful. Thomson's investigations opened the field of mass spectroscopy in which great success was achieved after the First World War by F. W. Aston (1877–1945) who developed spectroscopes that gave the ionic mass with greater and greater precision.

One of Aston's important results was that the relative weights of all the atoms examined, except hydrogen and lithium, were found to be a whole number to the accuracy of measurement, 1 part in a thousand, when referred to $\frac{1}{16}$ of the weight of O^{16} as unit. Nuclei were then assumed to be made out of protons and electrons. The protons conferred the mass; the electrons adjusted the charge and, being about $\frac{1}{1840}$ as heavy as the protons, did not greatly alter the nuclear mass. The divergences from the "whole number rule" were interpreted as due to the loss of mass occurring when free particles coalesce, according to Einstein's law $E = mc^2$. (See also p. 136.)

But let us return to Manchester. The atomic model, after it was formulated by Rutherford, was further developed more by Bohr than by Rutherford. Instead, Rutherford proceeded with the study of beta and gamma rays, obtaining worthy but not revolutionary results. Meanwhile, England had entered the First World War, and the population of the Manchester Laboratory diminished. Rutherford, too, became more and more involved in work for the Admiralty and for the defense of the Empire. These commitments took him to America, where he spent considerable time in Washington. Meanwhile, he had become Sir Ernest. It was a sign of the still relatively civilized times that during the war Rutherford was able to maintain a correspondence with Stefan Meyer in Austria and with Geiger in Germany. Geiger, in turn, arranged for J. Chadwick, one of Rutherford's best students, who was captured and interned in Germany during the war, to continue with research there.

The Disintegration of the Nucleus

By 1917 Rutherford was one of the very few remaining scientists in the Manchester Laboratory, along with his technician, Kay. However, he lacked able students to whom he could entrust experiments. Marsden, the faithful Marsden, before returning as a professor to his native New Zealand, in 1915 had observed a strange phenomenon, the presence of some particles with exceptionally long ranges when he bombarded air with alpha particles. A possible explanation was that they were hydrogen nuclei, because such long-range recoils appear when hydrogen is bombarded with alpha particles. But Rutherford suspected that it was something else of colossal import, and in a long and patient study carried out mainly when his official obligations left him time, he decided to verify the nature of the particle projected. In a paper in November 1917, he asked whether they were atoms of N, He, H, or Li.

By June 1919 Rutherford was ready to publish a paper entitled "Collisions of Alpha Particles with Light Atoms." The work was composed of four parts. The first three are excellent but more or less routine investigations, but the fourth, subtitled "An Anomalous Effect in Nitrogen," states:

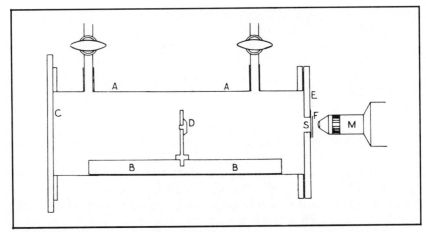

Figure 6.4 Apparatus used by Rutherford in observing the first nuclear disintegration. The illustration is taken from Rutherford, Chadwick, and Ellis, *Radiations from Radioactive Substances* (Cambridge University Press, 1931), where it actually appears twice! It shows an air-tight container that can be filled with gas (nitrogen) and contains a source of alpha particles placed in D. Since the range DS is greater than that of the alpha particles, one can conclude that particles causing the scintillations on the screen F are emitted in the disintegration of nitrogen gas nuclei hit by alpha particles. A detailed study shows that the fragments are protons.

> We must conclude that the nitrogen atom is disintegrated under the intense forces developed in a close collision with a swift alpha particle, and that the hydrogen atom which is liberated formed a constituent part of the nitrogen nucleus. ... The results as a whole suggest that if α particles—or similar projectiles—of still greater energy were available for experiment, we might expect to break down the nucleus structure of many of the lighter atoms. [*Philosophical Magazine* 37, 581 (1919)]

This was nuclear disintegration, the alchemists' dream in a modern form. Rutherford had carried out every possible control experiment before announcing his discovery. He wanted to be absolutely certain of his results, and that took him about three years. Figure 6.4 shows the apparatus Rutherford used. Rutherford's apparatus cost less than a millionth of what a modern accelerator costs; however, it needed *his* eye at the microscope, a requirement not easily fulfilled.

Rutherford's experiments were repeated in Vienna, and Austrian scientists found more disintegrations than Rutherford did. A lively debate arose, but in the end it was found that Rutherford was right. At the Cavendish Laboratory, P. M. S. Blackett obtained images in the Wilson cloud chamber that confirmed Rutherford's results (Figure 6.5).

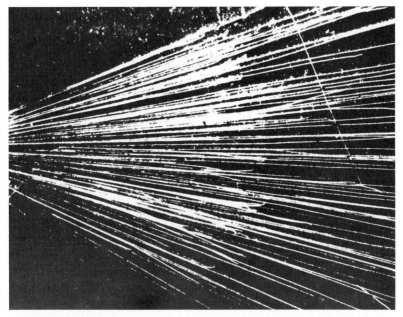

Figure 6.5 The disintegration of a nitrogen nucleus in a cloud chamber, as observed by Blackett. The source contains $Pb^{212} + Bi^{212} + Po^{212}$ in radioactive equilibrium and emits alpha particles with two ranges: 8.6 and 4.8 cm. One of the particles with a longer range hits a nitrogen nucleus and breaks it according to the reaction $_7N^{14} + {}_2He^4 = {}_8O^{17} + {}_1H^1$. The longer transverse trace is that of the proton, the other that of $_8O^{17}$ [From P. M. S. Blackett and D. Lea in *Proceedings of the Royal Society, London 136*, 325 (1932).]

Director of the Cavendish Laboratory

At the end of the war, J. J. Thomson retired from the Cavendish Laboratory (Figure 6.6) and became master of Trinity College at Cambridge. A worthy successor to the post of Cavendish Professor and director of the laboratory was sought, but it was not an easy task. The new director would have to follow an impressive line of succession: Maxwell, Rayleigh, and J. J. Thomson. In scientific stature the obvious candidate was Rutherford. The successor also had to have considerable self-confidence in view of possible comparisons with his predecessors. Rutherford had no fears and, with good reason, was not modest. However, he wanted to be sure that J. J. Thomson would not try to retain too much influence in the laboratory. With complete openness he wrote this to his former professor, and the two superior personalities clarified and settled this point, avoiding later disagreements.

Rutherford was invited to give the Bakerian Lecture for the second time in 1920. As on the first occasion (in 1904), he summarized the work he

Figure 6.6 The facade of the Cavendish Laboratory in Cambridge as it appeared from Maxwell's time onward. The building at present is devoted to other university uses and the Cavendish Laboratory has been relocated. (Cavendish Laboratory, Cambridge University.)

had carried out, this time dealing with the Manchester period from the formulation of the nuclear atom model to the disintegration of the nucleus. In describing the atom, he quoted details from his experiments on scattering alpha particles and then cited works of Barkla and Moseley that established the existence of the atomic number. He was still very reserved about Bohr's work. He then described in detail his disintegration work. In this lecture, he also expressed some tentative ideas about the possible existence of a neutral particle with a mass similar to that of the proton (a term he had coined to refer to the hydrogen nucleus). He considered this hypothetical particle rather like a hydrogen atom in which the electron had fallen inside the nucleus, neutralizing it electrically. This speculation proved to be important, as we will see later. He also speculated about a possible hydrogen isotope of mass 2 (deuterium).

At the Cavendish Laboratory Rutherford was no longer working in the old style, with his own hands. By now he was in his fifties and was surrounded by a new generation of young scientists who were returning from the war. His main responsibilities at Cambridge were to direct the laboratory and inspire and guide many excellent young physicists, including J. Chadwick, P. M. S. Blackett, C. D. Ellis, J. D. Cockcroft, E. T. S. Walton, M. Oliphant, C. E. Wynn-Williams, and others who were working on problems in nuclear physics (Figure 6.7).

In the same building or nearby there were other physicists who did not move in Rutherford's circle. Among them were J. J. Thomson, who had maintained his own laboratory, F. Aston, C. T. R. Wilson, and later the Russian P. Kapitza. Although Rutherford was not carrying out experiments himself, he followed what was happening in the laboratory and often provided ideas, including fine working details for anyone who needed them. But above all, he gave general direction to the laboratory and decided on lines of research.

I remember the atmosphere in 1934 when E. Amaldi and I spent a few weeks at the Cavendish Laboratory. Our work in Rome on neutrons procured us a warm reception, and Rutherford showed a lively interest in our results, questioning us on several points. We asked him to communicate our work to the Royal Society. He took the manuscript, and the following day brought it back with numerous corrections in his own handwriting, mainly for the purpose of improving our English. When I asked him if he could arrange for rapid publication, he laughed and promptly answered, "What do you think I was the President of the Royal Society for?"

When he went around the laboratory, he would sit on one of the laboratory stools, extract the butt of a pencil from his waistcoat, and check the results of the experiment in progress. The research workers, when addressed, nearly stood at attention and ran rather than walked. This response certainly did not arise out of formal discipline but from the intrinsic respect that Rutherford elicited. A comment from Rutherford, whether good or bad, was not taken lightly. In other places I have seen famous laboratory directors treated almost condescendingly by young scientists, but this certainly did not happen to Rutherford.

Figure 6.8, Rutherford and J. A. Ratcliffe, is slightly ironic, because it shows an amplifier for detecting alpha particles that could not tolerate any noise. For proper functioning it required silence or at least a subdued voice, which was not one of Rutherford's characteristics.

Rutherford's attitude toward theoretical physics was peculiar. He was certainly not a theoretician himself, and he was quick to make fun of theories and theoreticians. Yet he paid close attention to theoretical results and translated them into the concrete terms that he preferred: He went so far as to say jokingly that alpha particles were red, as anybody knew. Bohr, who greatly respected Rutherford, had been his protegé, and they undoubtedly must have discussed physics together, but their means of communication, with their differences, remains a mystery.

Though Rutherford was a superb teacher for research scientists who had the good fortune of being near him, his classroom lectures were not

outstanding. He would get embroiled in the subject he was discussing or easily digress, slipping into one of his favorite topics. Once in a lecture, when faced with an integral that should have canceled out but having forgotten the reason, he said with complete seriousness that it canceled out because the differential was infinitesimal.

Tradition reports innumerable anecdotes about Rutherford. Some of them can be found in the books by his friends A. S. Eve and Mark Oliphant, listed in the bibliography. I will relate only a few that help characterize Rutherford's exceptional personality.

Occasionally, waxing enthusiastic about science, Rutherford would

114

Figure 6.7 A group of research workers at the Cavendish Laboratory in June 1932. In the front row, from left to right: J. A. Ratcliffe, P. Kapitza, J. Chadwick, R. Ladenburg, J. J. Thomson, E. Rutherford, C. T. R. Wilson, F. W. Aston, C. D. Ellis, P. M. S. Blackett, J. D. Cockcroft. (Cavendish Laboratory, Cambridge University.)

say that his era was comparable, in intellectual vigor, with Elizabethan times, and he left no doubt as to who was the modern Shakespeare.

A famous philosopher and Rutherford were talking about their respective disciplines. Rutherford proclaimed that philosophy was nothing but hot air. Hot air! To which the philosopher replied that Rutherford was a savage. "A noble savage, I admit, but still a savage!" The philosopher then recounted a tale of Napoleon III's Marshal McMahon: "The Marshal was reviewing a regiment in which there was a Negro cadet, and he had been asked to say something encouraging to him. Having reached the Negro's platoon, the Marshal stopped, looking at the cadet, and then said to him,

Figure 6.8 Rutherford and Ratcliffe inside the Cavendish Laboratory, about 1932. The sign "Talk softly please" refers to the necessity of avoiding noise that would disturb the apparatus on the cart, the instrument for revealing alpha particles. (Photo by C. E. Wynn-Williams, from Eve, *Rutherford*.)

'Cadet, you are a Negro,' to which the cadet replied, 'Yes sir!' A long pause, and then the Marshal said, 'Well, go on being one.' And that's what I say to you, Rutherford, go on."

Mark Oliphant, one of Rutherford's last collaborators, relates that in using one of the first accelerators, they had found some particles coming

Figure 6.9 Ernest Rutherford (right) in his later years, with J. J. Thomson, his predecessor as director of the Cavendish Laboratory. Thomson survived Rutherford by a few years. (Cavendish Laboratory, Cambridge University.)

from the reaction of deuterium plus deuterium, but the nature of the particles was unclear. After a long day's work Oliphant went home to bed, but in the middle of the night he was awakened by the telephone. Oliphant's wife answered and, slightly alarmed on hearing Rutherford's voice, she called her husband. Lord Rutherford said, "I know what the particles are. They are helium of mass 3." The sleepy Oliphant answered right away, "Yes sir, but why do you think they are helium 3?" to which Rutherford replied, "Reasons, reasons! I feel it in my water." Naturally Rutherford was right, as Oliphant verified the next day.

According to Rutherford's close friend Kapitza, Rutherford's extraordinary intuition can be partly explained by the enormous intellectual activity he carried on. He formulated hypothesis after hypothesis, rejecting them or modifying them according to need, doing everything with inexhaustible energy. He worked all the time, and even his friends and colleagues barely knew a small fraction of his scientific thoughts. Sometimes from little hints they realized that he had attempted an experiment unsuccessfully. This explanation of his "intuition" seems valid to me and I am sure it also applies to other great scientists.

Rutherford had received the highest scientific honors from countries all over the world, and was President of the Royal Society from 1925 to 1930. On January 1, 1931, he was made a peer. He sent a telegram to his mother, then nearly ninety years old, who still lived in New Zealand, It said,

"Now Lord Rutherford; more your honor than mine, Ernest." The coat of arms of the new Baronet Nelson shows a stylized rendering of his decay and growth curves going back to his Canadian days (see Figure 3.6b). Politically, Rutherford was rather conservative, and not very active. But when Hitler began the persecution of the Jews, an Academic Assistance Council was formed in England with the aim of helping the victims of the Nazis, and Rutherford became its president.

In his final years (Figure 6.9) Rutherford saw changes in physics that perhaps did not suit his nature. Experiments were becoming more complicated, accelerators were born, and theory was becoming increasingly abstract. This new era of nuclear physics, which began in part at the Cavendish Laboratory, will be treated later. Although some of the protagonists were pupils of Rutherford, many came from diverse backgrounds and from a wide scientific circle.

In 1937, during a meeting commemorating Galvani held in Bologna, word arrived that Rutherford was seriously ill with a hernia. On October 19, 1937, he died. The death was announced at the meeting by Bohr, his voice broken by tears. Although many of the participants at the meeting knew Rutherford only from his scientific works, the expressions on people's faces showed what a great loss it was. He is buried in Westminster Abbey, close to the tomb of Newton.

Chapter 7

Bohr and Atomic Models

The young physicists surrounding Rutherford at Manchester were almost all experimentalists. Rutherford himself had mixed feelings toward theory: He was too intelligent to ignore its importance, but he thought intuitively, using simple models, following the English tradition. The immense success that resulted from his simple experimental and intellectual methods perhaps excessively reinforced his confidence in this approach. It seems that Rutherford had a very limited interest in quanta and in the great new ideas that were revolutionizing theoretical physics; he was intent on his own revolution. An exchange of ideas regarding physics between Rutherford and Einstein was hardly thinkable, at least from Rutherford's point of view. Nevertheless, it was in his laboratory at Manchester that the next theoretical revolution started, fomented by a visitor who participated very actively in the laboratory's life: Niels Bohr (1885–1962).

The Young Bohr and the Hydrogen Atom

Niels Bohr (Figure 7.1) was born on October 7, 1885, at Copenhagen, the son of the distinguished physiologist Christian Bohr and his wife Ellen Adler, the daughter of a wealthy Jewish banker. The family offered every advantage for a full academic and cultural education to Niels and his younger brother Harald, who became a famous mathematician. The two children were doted upon by their mother and her sisters. The Bohrs were an upper-middle-class Danish family, and in such a small country they had access to all the intellectually prominent individuals of the period; they especially associated with philosophers and medical men. There is no evidence that Niels was a prodigy, although he made some remarkably accurate drawings as a child. If I understood him correctly, Bohr once told me of the difficulty he experienced in learning to write.

Figure 7.1 Niels Bohr (1885–1962) about the time he was developing his atomic model, the Bohr atom. (Niels Bohr Institute).

Both Niels and his brother became remarkable athletes at a young age. They played soccer almost at a professional level, and I remember Bohr, well past his sixtieth birthday, deftly skiing on the slopes near Los Alamos. In high school the brothers started to advance beyond their classmates, and in 1904, when Niels was nineteen and his brother Harald was seventeen, a schoolmate already referred to them as geniuses.

Niels's interest in physics was first aroused by his father when the child was still in elementary school. When Niels had to choose a field at the University of Copenhagen, he decided in favor of physics and studied under Professor C. Christiansen. He worked on liquid jets and surface tension in order to participate in a competition held by the Danish Academy of Sciences in 1905. He won the prize with a theoretical and experimental investigation, performing the experiments in his father's laboratory. Later, however, he concentrated increasingly on theory. He wrote his doctoral dissertation on the electronic theory of metals and soon thereafter, in 1911, went to work under J. J. Thomson at the Cavendish Laboratory. Thomson received him courteously and listened to his explanations of his work, but could not find time to read his thesis. This is not too surprising in view of Thomson's own work schedule and the number of students he had to supervise. In Cambridge Bohr met Rutherford and was impressed by him to the extent that, in November 1911, Bohr went to Manchester to take an experimental course on radioactive measurements offered in Rutherford's laboratory. While waiting for a radioactive source to arrive, he first worked on a subject that was fashionable at Manchester: the passage of alpha particles through matter. He obtained some interesting results, and this remained one of his favorite subjects to the end of his life. Soon, however, he passed to an area of much greater importance, which also was connected with Rutherford's activities.

As we have seen, Rutherford had developed an atomic model that could account for the big deflections occasionally experienced by alpha particles in passing through matter. In order to account for this phenomenon Rutherford had considered a "Saturnian" model of the atom, or, in more modern language, a nuclear model. Bohr took this model very seriously in spite of the difficulty presented by its mechanical and electrical instability. This difficulty, already pointed out by Rutherford, had some interesting features. It was known or assumed that all atoms of a substance are equal, but there was nothing in the model that could assure this equality; in particular, there was nothing that could determine the atomic radius. Thus if the model was to survive, it was necessary to find a radical remedy that would give both stability and a fixed radius. The charge and mass of the electron are ingredients that would enter in any calculation; what could be another universal constant (independent of the substance) that could give a length?

The answer was not too recondite for a person familiar with the "modern" ideas of 1911. The quantum of action, Planck's constant, had to play a role. There were several possibilities. The view that atomic size was the fundamental constant and that the quantum of action followed from it had been put forward by the Austrian physicist A. Haas. The British astronomer J. W. Nicholson had also tried to introduce h into atomic models, and the Danish chemist N. Bjerrum had tried to introduce it into molecular models. However, these attempts had been vague or had gone in mistaken directions.

Bohr pondered these problems and ideas and mentioned them in a letter to his brother dated June 19, 1912. In June or July he prepared a memorandum for a discussion with Rutherford on the subject. He was obviously very enthusiastic about the model but had not yet considered the hydrogen spectrum. Spectra became the key to much that came later, but they were considered too complicated and a seemingly intractable field at the time. It was early in 1913 that a student friend, Hans Marius Hansen, asked Bohr what his model had to say about spectra. When Bohr said that he could not say anything on this subject, Hansen advised him to have a look at Balmer's formula. "As soon as I saw Balmer's formula the whole thing was immediately clear to me," Bohr said many years later.

What was Balmer's formula? Johann Jakob Balmer (1825–1898), a Swiss high school teacher and something of a numerologist, in 1885 had noted a striking regularity in the frequencies of the spectral lines of hydrogen. These are given by the formula

$$\nu = R \left(\frac{1}{n_1{}^2} - \frac{1}{n_2{}^2} \right)$$

where n_1 and n_2 are positive integral numbers, and $n_1 < n_2$. (I am using more modern notation than Balmer's; he used wavelengths in lieu of frequencies). The formula, which became a touchstone for any theory of the hydrogen atom, is obeyed with great accuracy by the line spectrum of hydrogen. At Balmer's time the visible series with $n_1 = 2$ was known, and this was the series in which he discovered the regularity (Figure 7.2). Later the ultraviolet series $n_1 = 1$ was found by T. Lyman, followed by the infrared series with $n_1 = 3$ (Paschen) and $n_1 = 4$ (Brackett). Figures 7.3 and 7.4 show these regularities graphically. The constant R, called the Rydberg constant in honor of the Swedish spectroscopist by this name, fixes the scale of the figure.

The orbits of a point attracted by an inverse square force to a fixed center are, as in the case of planets, ellipses with the fixed center in a focus, but for the sake of simplicity we shall consider only the special case of circular orbits with the fixed center at the center of the orbit. For such a system any radius is possible, provided the velocity in the orbit is such that the centrifugal force compensates exactly the attraction from the center. This occurs if the radius and the velocity satisfy the relation

$$\frac{mv^2}{r} = \frac{e^2}{r^2}$$

Thus any radius is possible provided we choose a suitable velocity; we must introduce some other conditions to fix the radius. Bohr gave criteria to choose among the infinite orbits that seem to be possible (Figure 7.5). These privileged orbits are called *stationary states*. Bohr postulated further, in his words:

> 1. That the dynamical equilibrium of the systems in the stationary states can be discussed by the help of the ordinary mechanics, while the passing of the system between different stationary states cannot be treated on that basis.

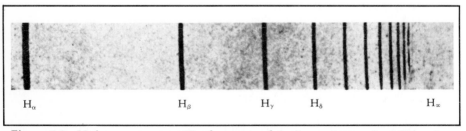

Figure 7.2 Hydrogen spectrum. The frequency of the lines corresponds to Balmer's formula. The lines become closer in frequency toward the limiting value H$_\infty$, after which the spectrum becomes continuous.

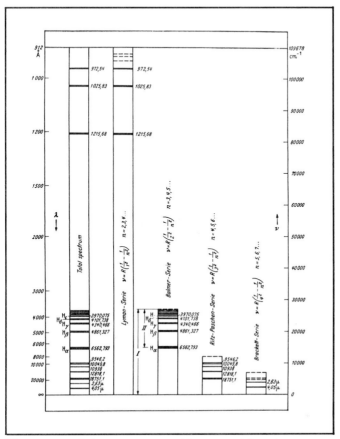

Figure 7.3 Hydrogen spectrum. The spectral lines have a frequency $\nu_{ij} = (E_i - E_j)/h$ where $E_i = R/n_i^2$ and $E_j = R/n_j^2$, where n_i and n_j are positive whole numbers and $n_i < n_j$. For n_i equal to 1, 2, 3, ..., one has the spectral series of Lyman, Balmer, Paschen, Brackett, Pfund, etc. R is equal to $2\pi^2 Z^2 e^4 m/h^3$, where m is the reduced mass. [From W. Grotrian, *Graphische Darstellung der Spektren* (Berlin, 1928).]

2. That the latter process is followed by the emission of a *homogeneous* radiation, for which the relation between frequency and the amount of energy emitted is the one given by Planck's theory (i.e., $E_1 - E_2 = h\nu$).

These two hypotheses contradict classical physics, and by postulating them simultaneously or separately, we enter a labyrinth from which it seems impossible to escape. Poincaré once observed that by postulating contradic-

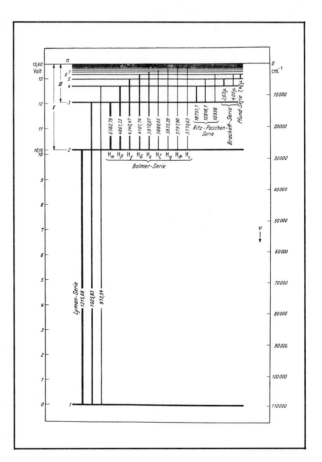

Figure 7.4 Energy levels of the hydrogen atom according to the Bohr model. The energies (scale on the left) are in electron volts; the "frequencies" are in wave numbers $1/\lambda$ (scale on the right). The true frequency is $c \times$ "frequency" in wave numbers since $\nu = c/\lambda$. [From W. Grotrian, *Graphische Darstellung der Spektren* (Berlin, 1928).]

tory hypotheses one can prove anything; this statement is mathematically correct. Only a rare and uncanny intuition saved Bohr from getting lost in the maze. Einstein, who was a master at guessing the most hidden natural truths, said many years later:

> That this insecure and contradictory foundation was sufficient to enable a man of Bohr's unique instinct and perceptiveness to discover the major laws of the spectral lines and of the electron shells of the atom as well as their significance for chemistry appeared to me like a miracle and appears as a miracle even today. This is the highest form of musicality in the sphere of thought. [Schilpp, *Albert Einstein, Philosopher-Scientist,* p. 46]

The Ariadne's thread that Bohr used for escaping from the labyrinth is the requirement that, for large orbits, classical theory and Bohr's hypothesis give the same result. In Bohr's own words,

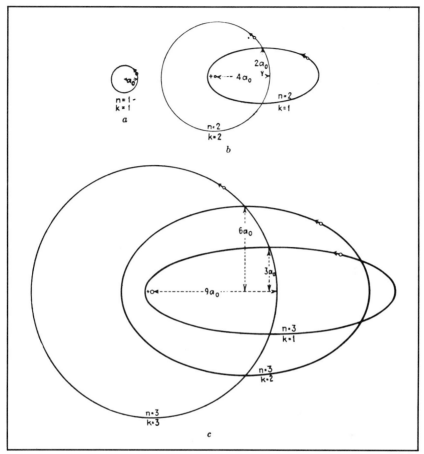

Figure 7.5 The electron orbits in the hydrogen atom according to Bohr and Sommerfeld's theory. The unit of length $a_0 = (h/2\pi e)^2(1/m)$ is the first Bohr radius and is 0.54×10^{-8} cm. n is called the total quantum number and determines the main axis of the orbits and the energy. The quantum number $k \leq n$ gives the eccentricity of the orbit; $k - 1 = l$ gives the angular momentum in $h/2\pi$ units.

We shall show that the conditions which will be used to determine the values of the energy in the stationary states are of such a type that the frequencies calculated ($E_1 - E_2 = h\nu$) in the limit where the motions in successive stationary states comparatively differ very little from each other, will tend to coincide with the frequencies to be expected on the ordinary theory of radiation from the motion of the system in the stationary states.

Revolution yes, but without irreparable destruction. This is a special case of the so-called correspondence principle, which shows how to move from macroscopic to microscopic systems. This principle was formulated in many ways and elaborated upon in all possible manners before the discovery of quantum mechanics. It is most useful as a guide to intuition, but it cannot be formulated rigorously as can, for instance, the principles of thermodynamics. Rather, it may be described, with some exaggeration, as a way of saying "Bohr would have proceeded in this way."

With the hypotheses formulated under (1) and (2) above, it is not difficult to obtain the radii of the orbits and the energy of the stationary states, and hence the spectrum of hydrogen (see Figures 7.4, 7.5).

In Appendix 9 we give the main points of Bohr's calculation as based on the correspondence principle. The final results are the orbital radii (r_n) increasing as the numbers 1, 4, 9, etc., or, more precisely,

$$r_n = \frac{n^2 h^2}{4\pi^2 m Z e^2}$$

where n is a positive integral number, h Planck's constant, m and e the mass and charge of the electron, and Z a positive integer (atomic number) such that Ze is the nuclear charge with the sign reversed. The energies of the stationary states are then

$$E_n = -\frac{Ze^2}{r_n} = -\frac{2\pi^2 Z^2 m e^4}{h^2 n^2}$$

where the energy is taken to be zero when the electron is at rest, at great distance from the nucleus. From the energy levels the frequencies of the spectral lines follow at once from rule (2), which is the relation

$$\nu_{nm} = \frac{E_n - E_m}{h}.$$

We have thus explained Balmer's formula, and the constant that appears in it can now be expressed through e, m, and h: It is

$$R = \frac{2\pi^2 Z^2 m e^3}{h^3}$$

Thus we have a bridge between chapters in physics that are very remote from each other: spectroscopy, radioactivity, and blackbody theory. This uniting of such apparently distant fields is always a sign that we are on the right path and is in itself a most important result. Figures 7.3, 7.4, and 7.5 show all this graphically: the energy levels, the line spectra, and the orbits.

Early in 1913 Bohr started writing a formidable trilogy of papers. In the first he expected to give a general explanation of the constitution of atoms and molecules. The second was to have been devoted to systems with several degrees of freedom, and the third was to have treated systems with more than one nucleus. The project was never completely finished, and only part of it appeared around 1918. In August 1918 Bohr wrote to his friend, the physicist O. W. Richardson, "I know that you understand how things happen and how my life, from the scientific point of view, passes periods of overhappiness and despair, of feeling vigorous and overworked, of starting papers and not getting them published, because all the time I am gradually changing my views about this terrible riddle which the quantum theory is. ..." [Bohr, *Collected Works,* vol. 3, p. 14]

The first atomic paper by Bohr, dated April 5, 1913, and published in volume 26 of the *Philosophical Magazine,* is easy to read for anybody with a slight knowledge of physics. The sophisticated reader will admire the dexterity with which Bohr sails across a sea full of treacherous shoals and lands safely under the guidance of the correspondence principle. At the end of the paper, when he is already in port, he adds a remark of the greatest importance: "For the allowed orbits, the angular momentum is an integral multiple of $h/2\pi$ and this fact may be also used as a quantization criterion, yielding the stationary states."

Bohr's theory demonstrates extraordinary power in predicting new spectral series besides Balmer's, in predicting the spectrum of ionized helium, in predicting the effect on the spectral lines of the finite nuclear mass, and so on. However, Bohr fully realized its grave imperfections and that, at best, it could only represent a temporary stage before a consistent theory was found. He always emphasized this disagreeable truth. Nevertheless, theoretical and experimental results, sometimes only approximate, yet truly impressive, kept accumulating. I will mention only two results, outside of spectroscopy, that touch upon the foundations of physics. These were the Franck-Hertz experiment and the Stern-Gerlach experiment, which I will treat later.

How was Bohr's theory received? Ex post facto testimony is not always reliable, but the fact that Rutherford communicated Bohr's paper to the *Philosophical Magazine,* despite its "wild ideas," shows that he thought it had merit. Bohr had sent him the manuscript from Denmark, and Rutherford, with his common sense, objected that in Bohr's theory an electron had to know beforehand which orbit it wanted to jump to. This serious difficulty was not clarified for some time. The decisive step was taken by Einstein in 1917 when, as we have seen, he introduced probability concepts based on the analogy with radioactivity. In any case, apart from this objection, Rutherford found the paper a little too long and tried to shorten it. He wrote somewhat casually to Bohr about doing so; the reaction was quite unexpected: Bohr took a boat to England and went to Manchester, where he defended his paper against Rutherford's criticism line by line, until he carried the day. I should like to have witnessed the scene. Bohr was a mild-mannered person and was courteous and respectful toward Rutherford. Rutherford was older than Bohr, was accustomed to giving orders, and had an established and great scientific position. Rutherford commented in a letter that he would never have expected such obstinacy from Bohr.

To give some idea of the causes of Rutherford's criticism, here is the concluding paragraph of one section of Bohr's first paper:

> The preliminary and hypothetical character of the above considerations need not to be emphasized. The intention, however, has been to show that the sketched generalization of the theory of the stationary states possibly may afford a simple basis of representing a number of experimental facts which can not be explained by help of the ordinary electrodynamics, and that the assumptions used do not seem to be inconsistent with experiments on phenomena for which a satisfactory explanation has been given by the classical dynamics and the wave theory of light.

The style may be stilted, but every word is thought through and carefully weighed. Needless to say, Rutherford and Bohr remained very close friends for life. Although they were very different in scientific style and approach, each had the highest regard and affection for the other, and they could understand each other scientifically. An eloquent testimonial to their relation is given by Bohr in his "Reminiscences of the founder of nuclear science and of some developments based on his work."

Writing was a painful experience for Bohr, as his later collaborators learned. A paper was started, written, and rewritten maybe up to a dozen times. In each version he changed a phrase here or there, clarified a statement, or improved the definition of a concept. At the end, it was sent for publication, to the Danish Academy of Sciences, for example. When the proofs were returned to the author, they were practically rewritten to increase the precision of expression or because some new idea had occurred to Bohr. The printer fretted and the academic authorities were embarrassed, but they tolerated the procedure because Bohr was Bohr. After several months of delay the volume of the proceedings would appear without Bohr's article, and a few years later the paper would still be in progress, although co-workers and colleagues had seen various provisional versions in the interim. If and when the paper finally appeared, it was not likely to be a model of clarity even if it was very profound. Regarding profundity and clarity of papers, Bohr had ready on hand certain German quotations such as the one on p. 170. This is perhaps more characteristic of the older man than of the thirty-year-old Bohr, but the letter to Richardson quoted on p. 127 and the Bohr archives bear ample testimony that even young Bohr had written this way.

Let us return to the reception accorded to the great 1913 paper. In September of that year a meeting of the British Association was held in Birmingham, attended by Rutherford, H. A. Lorentz, O. Lodge, Lord Rayleigh, Jeans, and other important scientists. Bohr's papers were noted and discussed. Lord Rayleigh commented, "I have looked at it, but I saw it was no use to me. I do not say that discoveries may not be made in that sort of way. I think very likely they may be. But it does not suit me" [Rayleigh, *Life of Lord Rayleigh,* p. 357]. Rayleigh was seventy-one years old and had long since vowed to keep his mouth shut about new developments in physics. He kept his promise. However, the success of the new theory could not be denied. At Göttingen, P. Debye was enthusiastic, and apparently A. Sommerfeld remarked to L. Brillouin that it was a paper of historical importance. On the other hand, in Zurich O. Stern and M. v. Laue, who later were to do so much for the advancement of Bohr's ideas, studied the paper and said that if by chance it should prove correct, they would quit physics. George Hevesy, who had become a close friend of Bohr at Manchester, visited Einstein and told him of Bohr's work. Einstein was immediately enthusiastic; he said that he had had similar ideas once but did not dare to pursue them. With obvious satisfaction Hevesy rushed to report Einstein's opinion in letters to both Rutherford and Bohr.

Bohr's atom preceded World War I by only a few months. Denmark was able to remain neutral, but Bohr sympathized with the cause of the Allied Powers. He had a minor but rather unsatisfactory job in Copenhagen, and in 1916 he took a position in Rutherford's laboratory at Manchester until 1919, when he was recalled to Denmark and offered a professorship of theoretical physics. Bohr had always maintained close contact with experimental physics, and he endeavored to obtain a laboratory in association with his chair. After the war he succeeded, and a new institute was built at Copenhagen. Initially, Bohr lived at the institute. Later, in 1932, the Bohrs were the first family to be invited to occupy "The House of Honor," a mansion built by J. C. Jacobsen, the founder of the famous Carlsberg Breweries.

The house is now the semi-official residence for Denmark's greatest citizens in either the arts or sciences. It is a princely home, for which the founder left funds for upkeep. The house is near the brewery, built in neoclassical style, adapted however to the Danish climate; it also has a beautiful garden. Mrs. Bohr was a polished hostess, and over the years the Bohrs invited many important people, including some royalty, several prime ministers, and many great scientists, including Rutherford. They also welcomed a stream of bright young physicists who were working in or visiting Copenhagen.

Bohr's reputation spread quickly, and soon he started receiving invitations to lecture all over the world. In 1920 he went to Germany and for the first time met Planck, Einstein, and other major German theoreticians. In Holland he met H. A. Lorentz and Paul Ehrenfest, with whim he established an immediate and firm friendship. Einstein, too, liked Bohr; he wrote to Bohr after his visit to Berlin, "Not often in life has a person, by his mere presence, given me such joy as you. I understand now why Ehrenfest is so fond of you. I am now studying your great papers and in doing so—when I get stuck somewhere—I have the pleasure of seeing your youthful face before me, smiling and explaining ..." [Bohr, *Collected Works,* vol. 3, pp. 22, 634].

Bohr and his institute at Copenhagen also fulfilled a noble and necessary function that arose from the consequences of the war. The war had left a legacy of hatred and some childish but dangerous desires for revenge and vendettas. These feelings extended to some scientists. Those on the Allied side wanted to exclude their German colleagues from international scientific life; on the other hand, the Germans acted, unjustifiably, as victims. This type of behavior prevailed more in the official scientific world composed of older and often inactive scientists than among the younger ones, who were intent on their research. Bohr tried to fight the excesses and to reestablish good relations among the divided factions. His personality and the fact that he was a citizen of a neutral country helped him in this enterprise.

Bohr's scientific activity took a peculiar form. His working habits often consisted of thinking aloud and carrying on interminable discussions, for which he needed a partner. It was a sort of Socratic method, by which he developed his ideas while talking. Bohr's first assistant was the Dutchman H. A. Kramers. He was followed by many others, who later became famous. The Englishmen P. A. M. Dirac and N. F. Mott; the German W. Heisenberg; the Austrian W. Pauli; the Russians G. Gamow and L. D. Landau; the Americans J. C. Slater, H. Urey, and J. R. Oppenheimer; the Japanese Y. Nishina; and the Scandinavians O. Klein and S. Rosseland are among those who stayed for extended periods in Copenhagen. Dirac wrote of his conversations with Bohr:

> Very often I was just his audience during the process of thinking aloud. I admired Bohr very much. He seemed to be the deepest thinker I ever met. His thoughts were of a kind which were, I would say, rather philosophical. I did not understand them completely, although I struggled as hard as I could to understand them. My own line of thinking was really to put emphasis on thoughts which could be expressed in the form of equations, and much of Bohr's thoughts were of a more general character and rather remote from mathematics. But still I was very happy to have this close connection with Bohr and, as I mentioned once before, I am not sure to what extent hearing all these thoughts of Bohr influenced my own work. [Dirac, *Proc. of the Intl. School of Physics,* "Enrico Fermi," vol. 57, p. 134]

Because of his peculiar way of writing, Bohr himself did not publish much; nevertheless, he was the center of all the activity. Furthermore, he traveled frequently and was invited to communicate his ideas in various scientific centers in Europe and in the United States.

Bohr was much more cautious and critical of the type of quantum mechanics based on models that existed before 1924 than other distinguished physicists, such as Arnold Sommerfeld. He always stressed the need to put the theory on solid foundations and emphasized the prevailing contradictions.

In Copenhagen Bohr also initiated an annual series of small private conferences at his institute. He invited about thirty physicists, choosing

Figure 7.6 Participants in the meeting organized by Bohr in Copenhagen in 1937. In the front row, from left to right: N. Bohr, W. Heisenberg, W. Pauli, O. Stern, L. Meitner, R. Ladenburg, J. C. Jacobsen. In the second row, seated from the left: V. Weisskopff, C. Moller, H. Euler, R. Peierls, F. Hund, M. Goldhaber, W. Heitler, E. Segrè. ... In the third row, seated from the left, G. Placzek, C. von Weizsacker, H. Kopfermann. ... Standing, H. D. Jensen, L. Rosenfeld, G. C. Wick. (Courtesy of E. Segrè.)

some famous ones, including Heisenberg, Pauli, and Stern, who attended practically all these meetings. In addition, he invited a number of active younger scientists from all countries. This was an opportunity to meet promising young physicists, to acquaint them with Copenhagen and with each other. The conferences often served to establish long-lasting friendships among physicists of the newer generations. There were few important physicists active between 1920 and the Second World War who were not given the unforgettable experience of participating in one or more of these conferences (Figure 7.6).

X-rays Come Into Their Own

Even if the fundamental themes of physics were voiced shortly before the First World War by Planck, Einstein, Rutherford, and Bohr, we should not forget the orchestration that accompanied them and greatly enhanced the music.

For over a century crystal structure had been attributed to a regular disposition of the constituent atoms. R. J. Haüy (1743–1822) had drawn

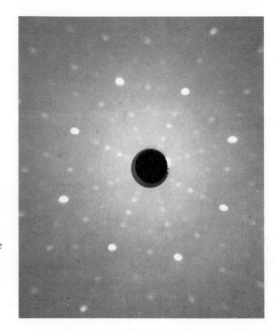

Figure 7.7 Diffraction picture of a monocrystal of tungsten obtained using the method of von Laue, Friedrich, and Knipping. (Courtesy of the CNR-LAMEL Laboratory, Bologna.)

beautiful pictures of the hypothetical atomic arrangements. Their modern versions are called space lattices and, as an example, we must imagine all atoms located at points having integral numbers as coordinates. Many of the laws of crystallography could be justified by admitting such hypothetical structures, but it was only in 1912 that they were really seen or demonstrated.

At Sommerfeld's Institute, in Munich, P. Ewald was studying the propagation of electromagnetic waves in a space lattice for his doctoral dissertation. At that time, around 1911, ideas about x-rays were developing and expanding. The hypothesis that they are short-wave electromagnetic vibrations similar to light seemed plausible, and estimates of their wavelength were being made. Some diffraction phenomena brought about by passing x-rays through a narrow slit could be interpreted by attributing to them a wavelength comparable to interatomic distances in a lattice.

Max von Laue (1879–1960) then did a theoretical study on the diffraction phenomena to be expected if one passed x-rays through a crystal. His results were immediately tested by W. Friedrich and P. Knipping with apparatus available in Röntgen's laboratory. They soon obtained excellent diffraction images that entirely confirmed von Laue's calculations (Figures 7.7, 7.8). The crystal acts as a three-dimensional diffraction grating, whose periodicity is given by the regularity of the atomic arrangement. Such natural gratings are about a thousand times finer than artificial gratings

Figure 7.8 Max Planck (left) and Max von Laue (right) during an excursion on Lake Como organized for the participants in the 1927 Volta Conference. (Courtesy of F. Rasetti.)

ruled, in their day, by Rowland, Michelson, and others, and they fit x-rays as well as the artificially ruled gratings fit visible light.

Von Laue's discovery has had innumerable ramifications. Almost immediately W. H. Bragg and W. L. Bragg, father and son, at Leeds, England used crystals as gratings to build a true x-ray spectrograph and started to extend the study of spectral lines to a region of wavelengths about a thousand times shorter than visible light.

Visible spectra originate from the motions of the outer atomic electrons, whereas x-ray spectra originate from the innermost. As a consequence, x-ray spectra show regularities unknown in optical spectra. They are dominated by the atomic number, that is, by the nuclear charge, as was shown by H. G. J. Moseley (1887–1915) immediately after the Braggs' discovery.

Moseley's story is poignant and tragic. He was descended from scientific families on both the paternal and maternal sides. He was educated at Eton in the best English public school tradition. At Manchester University he initiated his research work under Rutherford, but later he transferred to Oxford. He was obviously a major rising star among British scientists and was an extremely hard worker (Figure 7.9). In an experimental career of very few years he obtained results of permanent value and everlasting fame. At the beginning of the First World War he enlisted as a volunteer and, in spite of attempts by Rutherford and others to protect him from mortal

Figure 7.9 H. G. J. Moseley (1887–1915) in one of the few existing portraits of him. He was killed in action during the English expedition to the Dardanelles at age twenty-seven. His discovery of the atomic number, the integral number measuring the positive charge of the nucleus in units of the protonic charge, gave a definitive clarification of the concept of a chemical element. (University of Manchester.)

danger in order to save him for English science, he insisted on doing combat duty. He was killed in action at age twenty-seven in the Dardanelles expedition conceived by Churchill.

Before the war, Moseley had shown that x-rays give a simple method for measuring the nuclear charge Z, alias the atomic number. Certain x-ray lines prominent in the spectra have frequencies proportional to Z^2. Hence if one plots the square root of the reciprocal of the wavelength versus the atomic number, one obtains a straight line (Figure 7.10). If the x-ray spectrum of an element was lacking, it could easily be predicted by interpolation. In this way, in an afternoon, Moseley could solve the problem that had baffled chemists for many decades and establish the true number of possible rare earths. When the famous French chemist Urbain brought him rare earth samples on which he had labored for years, Moseley could analyze them in a few hours and reveal their content to the amazed chemist. Moseley's discovery is directly linked with Bohr's atom as Moseley recognized at once. It suffices to insert into the general formula giving the hydrogen spectral lines $n_1 = 1$ and $n_2 = 2$ to obtain Moseley's law, with the constant of proportionality specified.

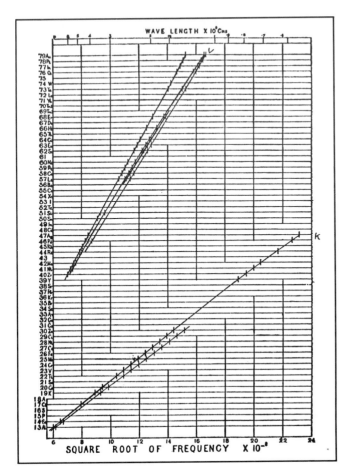

Figure 7.10 A Moseley diagram. The abscissa shows the square root of the frequency of the characteristic x-radiation emitted. The ordinate gives the atomic number. For the K line the frequency is given with a good approximation by the formula $\frac{3}{4}RZ^2$. [From *Philosophical Magazine 27*, 703 (1914).]

Moseley's discovery also provided a method for establishing the missing chemical elements with certainty. At his time they had the atomic numbers 43, 61, 72, 75, 85, and 87; and the periodic system ended with $Z = 92$ (uranium). In subsequent years element 72 was discovered in zirconium ores at Bohr's laboratory by Hevesy and D. Coster in 1923; they called it *hafnium*, after the latin name of the city of Copenhagen. Element 75 was discovered in 1925 by Ida and Walter Noddack in Berlin in several minerals, and in a burst of German patriotism they called it *rhenium* (the Rhine was a long-standing cause of friction between Germany and France). In these discoveries x-ray analysis was of paramount importance.

Element 87, francium, was found by Marguerite Perey as a rare branching in a natural radioactive family in 1939; and 91, protoactinium, was found by Otto Hahn and Lise Meitner in 1917 as a decay product of uranium. Radioactive methods were more important than x-ray analysis in the last two cases.

The remaining elements do not exist naturally on earth because they are radioactive and have relatively short half-lives; thus they decayed during geological times even if they were present at one time. They had to be formed artificially by nuclear bombardments. The first to be obtained in this way was element 43, technetium (meaning artificial in Greek). It was discovered in 1937 by Carlo Perrier and Segrè in molybdenum bombarded in the Berkeley cyclotron. Element 61, promethium, was prepared artificially by Charles Coryell and others in nuclear reactors in 1946. Element 85, astatine, was also prepared with the cyclotron by Corson, Mackenzie, and Segrè in 1940. The extension of the periodic system beyond uranium followed later, after Hahn and Strassmann's discovery of nuclear fission.

The new science of x-ray spectroscopy not only allows the study of deep electron shells and elementary chemical analysis on an unprecedented level of sensitivity and certainty; it also opens the way to the exploration of crystalline lattices and, more generally, the architecture of solids and of molecules. This technique has led to the new and active science of structural analysis, which is fundamental to many diverse sciences, ranging from mineralogy to molecular biology.

In fact, molecular biology has brought about a revolution in biology comparable to that caused by quantum mechanics in physics; it has opened entirely new horizons. This new branch of science has received indispensable help from modern physics through the use of radioactive tracers mentioned on page 108 and of x-ray structural analysis.

We have seen that the concept of isotopy had slowly become established for radioactive nuclei and that J. J. Thomson extended it to stable nuclei, proving the existence of two species of neon atoms, identical in everything chemical except atomic weight. Immediately after the war F. W. Aston started building a series of precision mass spectrographs that measured masses and abundances of isotopes. For many years Aston reigned supreme in this type of investigation, which provided many interesting results. We have mentioned the "whole number rule." Precise mass measurements coupled with Einstein's formula $E = mc^2$ gave information on the energy released or absorbed in nuclear reactions and thus became vital to nuclear physics. Today a combination of data from mass spectrography and nuclear reactions allows the establishment of nuclear masses with a precision of one part in 10^9 in favorable cases. The study of isotopes, however, has spread far afield and has had innumerable applications ranging from geology to archaeology and from vacuum technology to biology.

These are notable examples of the interdependence of different sciences. It is almost like observing a living organism, in which various organs cooperate in complicated and sometimes inscrutable ways.

The Quantized Atom Established

After Bohr's fundamental work, atoms and molecules moved to the center of attention for both experimental and theoretical physicists. Other areas of physics were still investigated by distinguished scientists, such as Rutherford, who never abandoned nuclear studies, or the Leyden experimental physicists, who concentrated on low temperatures, but atoms and molecules were the focus of most activity. Before Bohr spectroscopy was an almost empirical topic that did not go much beyond the cataloging of many spectral lines and the observation of conditions under which they were produced. The study of electric discharges in gases, which went back to Faraday, was also chiefly empirical. The new atomic theory provided a guide to the understanding of many phenomena and to the prediction of new ones; theory and experiment started to go hand in hand and at great speed. The center for such work was in Germany. One of the most significant steps was made before the First World War by James Franck and Gustav Hertz (a nephew of Heinrich Hertz of the electromagnetic waves). With conclusive experiments they showed the existence of the stationary states postulated by Bohr. To this effect they produced jumps between them, supplying the excitation energy by collisions with accelerated electrons. They observed the energy loss of the electrons that occurred only when the electrons had sufficient energy to push the target atom in an excited state. Furthermore, they saw the bombarded vapor emitting the spectral lines corresponding to the return from the excited state to lower states. The frequencies emitted obeyed the fundamental law

$$h\nu = E_1 - E_2$$

postulated by Bohr. These experiments tangibly proved some of the strangest of Bohr's hypotheses.

Shortly after the end of the war, in 1921, another experiment by Otto Stern and Walter Gerlach demonstrated one more fact required by Bohr's theory; it was a fact that seemed contrary to common sense, or better, to common macroscopic sensory experience. It is called space quantization (Figure 7.11). We have seen that Bohr's quantum condition may be formulated by saying that orbits must have an angular momentum that is an integral multiple of $h/2\pi$ which is $lh/2\pi$. Refinements of the quantum conditions discovered by Bohr, Sommerfeld, and others require the quantization not only of angular momentum but also of its component in a direction defined, for instance, by the presence of a magnetic field. This component may only have the values $mh/2\pi$, where m is a positive or negative integral number that is equal to or smaller in magnitude than l. It follows from this that an atom can have only certain orientations in space. For instance, if $l = 1$, m can have the values -1, 0, or 1, and the atom can be oriented in only

Figure 7.11 The result of the experiment by Stern and Gerlach in 1921 on space quantization. The two dark spots are caused by atoms (in a beam of lithium atoms) having opposite orientation in a magnetic field.

three positions with the angular momentum parallel, perpendicular, or antiparallel to the magnetic field. This extraordinary fact can be verified experimentally by sending a molecular beam, a tenuous stream of molecules or atoms, through a suitable magnetic field and observing their deflection. The deflection is produced by the action of a nonhomogeneous magnetic field on the magnetic moment of the atom, associated with the angular momentum. The atoms orient themselves only in discontinuous positions.

Otto Stern (1888–1969), who planned the experiment on space quantization, was the same man we encountered earlier as a follower of Einstein in Prague and Zurich, where he had vowed to abandon physics if Bohr's ideas were true. I consider him one of the major physicists of the century (Figures 7.12, 7.13). Starting in 1920, he devoted himself to the development of the molecular beam method. With this method a tenuous stream of molecules is produced in a high vacuum and, in this free state, they are subjected to electric or magnetic fields or to whatever one wishes to investigate. The importance of the method derives from the fact that the molecules are free and the experimental conditions closely approach those postulated in most theoretical treatments.

The French physicist L. Dunoyer was the first to produce a molecular beam in 1910. However, the method was developed and used chiefly by Stern, his pupils, and his co-workers. They investigated the fundamental assumptions of gas kinetic theory and in particular the Maxwellian distribution of velocities, the space quantization, the magnetic moment of many atoms, the magnetic moment of the proton, the de Broglie relation between momentum and wavelength for helium atoms, and so on. Stern, too, was exiled, and his institute was practically dismantled by the Nazis. He resettled later in the United States, but his most important work was concluded before his exile. Stern's tradition was continued and extended by the introduction of advanced radiofrequency and vacuum techniques, by I. I. Rabi

Figure 7.12 Otto Stern, one of the great experimental physicists of the period between the two world wars. He was responsible for classical experiments on space quantization, on de Broglie waves, and on the magnetic moment of the proton. These were all carried out using the molecular beam method. (University of Hamburg.)

Figure 7.13 Otto Stern having a discussion with the American physicist L. Langmuir during an excursion on Lake Como organized for the participants in the Volta Conference in 1927. (Courtesy of F. Rasetti.)

(1898–) and his school, mainly at Columbia University in New York City. They were able to improve immensely on previous measurements by Stern and to tackle entirely new problems. Rabi had been in Stern's 1930 laboratory as a postdoctoral Fellow. He had been impressed by the power of Stern's methods, but had not performed experiments in Stern's laboratory.

The Zeeman effect, by then about twenty years old, continued to be eagerly studied, especially in Germany, by F. Paschen, E. Back, and others. It was a steady source of precise and important results that greatly helped to unravel the structure of more complex atoms. The explanation of the Zeeman triplet given in classical terms by H. A. Lorentz could be translated into quantum terms and connected to space quantization. However, many atoms show much more complicated Zeeman patterns than the Lorentz triplet. They are called *anomalous Zeeman effects.* These presented a true challenge to atomic theory.

Alfred Landé first in Munich and later in Tübingen found semiempirical formulae that described these effects very accurately, but his results were very hard to understand in any model. For instance, in atomic mechanics, based on the correspondence principle, we could expect the square of the angular momentum in units $h/2\pi$ to enter certain formulae. We already know the angular momentum is a whole number l, but the formulae agreed with experimental data only if l^2 was replaced with $l\,(l + 1)$. When l is large, the two expressions coincide, but when l is small, they differ radically. Furthermore, angular momenta multiples of $h/4\pi$ and not of $h/2\pi$ were often found. In other words, half quanta seemed to appear. Landé's formulae agreed with experimental results, but their justification was a mystery to such scientists as Pauli. A friend of Pauli's saw him sitting on a park bench in Copenhagen looking dejected and asked what was making him so unhappy. Pauli answered, "How can one avoid despondency if one thinks of the anomalous Zeeman effect?"

To explain $l(l + 1)$, quantum mechanics was necessary; to explain the half quanta, the spinning electron was needed. Both concepts were just around the corner. The spinning electron, in its original formulation, is simpler and more intuitive than quantum mechanics, which will be treated in the next chapter.

In November 1925 George E. Uhlenbeck (b. 1900) and Sam A. Goudsmit (1902–1978), two Dutch physicists in their mid-twenties who were working at Leyden, made a most important discovery: The electron, which until then had been considered as a point electric charge, or perhaps as a little charged sphere, was found to have a spin, an intrinsic angular momentum. In other words, it turns on itself as the earth does. In addition, the electron has an intrinsic magnetic moment associated with its spin. The spin, as we might expect, is connected with the natural unit of angular momentum $h/2\pi$, but there is a novelty: It is not an integral multiple of such a unit, but half of it. At the same time, the magnetic moment associated with the spin is equal to the magnetic moment associated with one unit of *orbital* angular momentum. This unit is $eh/4\pi mc$, and is called the Bohr magneton. The ratio between angular momentum and magnetic moment for the spin is twice as large as for the orbits. Uhlenbeck and Goudsmit concluded all this from a study of atomic spectra.

Uhlenbeck had spent some time in Italy as a tutor to the son of the Dutch Ambassador, and there he had met and befriended Fermi. Both Uhlenbeck and Goudsmit were studying with Ehrenfest in Leyden when the idea of the spinning electron occurred to them. R. de L. Kronig, an American of Hungarian descent and another friend of Fermi, with whom he had spent some time in the Italian Dolomites, independently had the same idea at about the same time as Uhlenbeck and Goudsmit. Unfortunately for him, he asked Pauli's opinion on the subject. Pauli was already famous for his achievements and also for his critical insight, but this time he erred grievously and convinced Kronig that his hypothesis had no foundation. When Uhlenbeck and Goudsmit heard Pauli's critique, it seemed to them justified at first sight, and they wanted to withdraw their paper, which Ehrenfest had already sent to a journal for publication. However, Ehrenfest countered that they were so young that they could afford an unsound paper. Thus the article was not withdrawn, and it appeared in physics journals. Later Pauli's criticism was proved, chiefly by H. L. Thomas, to be unjustified. Pauli had objected that a certain factor of 2 in the formulae seemed to destroy the agreement with experimental evidence; Thomas showed that a subtle relativistic argument provided the missing factor. Thus Uhlenbeck and Goudsmit can justly be considered the discoverers of the electron spin, a major advance in physics. Unfortunately, they did not get a well-deserved Nobel Prize for it.

Weimar and Copenhagen Physics. The Exclusion Principle

Bohr's early success was mainly with the hydrogen atom, but from the very beginning he worried about atoms with more than one electron. The mechanical problems become immensely more complicated as soon as one passes from a two-body to a multibody system. The quantization rules had to be generalized, and the complicated hypothetical orbits arising had to be calculated. The generalization of the quantization rules was carried out chiefly by Arnold Sommerfeld (1868–1951), professor of theoretical physics at Munich. Originally Sommerfeld was a pure mathematician, and he had worked with Felix Klein at Göttingen. Klein was convinced that the relationship between mathematics and technical disciplines needed strengthening, and he converted Sommerfeld to his ideas. Sommerfeld changed his interest from pure mathematics and became a first-rate applied mathematician, bringing powerful mathematical techniques to engineering problems such as lubrication and the propagation of electromagnetic waves.

In 1906 Sommerfeld was called to the University of Munich as professor of theoretical physics, and there he established a major center for this discipline. He attracted many gifted students and had as his assistant Peter Debye, who helped arouse his interest in quantum theory. Sommerfeld had already been converted to relativity a little earlier. Under Sommerfeld's leadership theoretical physics thrived at Munich. His students acquired mathematical facility and learned up-to-date physics. Sommerfeld's influence was pervasive, and for over forty years he had a succession of students in theoretical physics whose brilliance was comparable to that of Rutherford's experimentalists. In his own studies Sommerfeld applied relativity to Bohr's atom, introducing the "fine structure constant" $2\pi e^2/hc$, a pure number that plays a very important role in physics. Its value, now known with great precision, is $1/137.03596$. There have been many attempts to derive it from bold comprehensive theories, but up to now these have been unsuccessful. As Sommerfeld said in his famous text "Spectral Lines and Atomic Constitution," on which a generation of physicists learned the subject (see p. 201), "In the fine structure constant e is the representative of the electron theory, h the appropriate representative of the quantum theory, c comes from relativity and characterizes it in contrast to classical theory."

In many ways Sommerfeld represented the German professor at his best. Behind his somewhat formal and rigid demeanor was a warm personality that generously cared for young students. Sommerfeld was a very good skier and had a house in the Bavarian Alps where he went during the winter with his students. On these occasions he lost even the appearance of a *Geheimrat,* and he coupled the art of skiing with discussions on physics in a most informal way. In fact, his interest in promising young physicists extended well beyond Germany. When Hitler came to power, Sommerfeld, who was a great German patriot, was not fooled by the Führer. I remember meeting him in 1934 in Holland, where he had gone to give a lecture. On that occasion he received a substantial fee, and he commented to me with great glee: "I am going to send it immediately to Rutherford for the benefit of displaced scholars. I cannot do so from Germany, but this is an opportunity I will not miss."

The methods of Bohr and Sommerfeld were applied by many physicists to phenomena such as the electric splitting of spectral lines, discovered by Johannes Stark in 1913, a phenomenon similar to the magnetic Zeeman effect. The methods were also applied to the detailed unraveling of x-ray spectra, a subject in which W. Kossel made decisive advances, and to the calculation of the intensity of spectral lines. Much of this work was influenced by Sommerfeld, from Munich, even if it was not carried out there. The results were always qualitatively correct, but quantitatively they were often only approximate. Again, the formulae often agreed with experiment when l^2 was replaced with $l(l + 1)$. All these considerations and attempts were guided by the correspondence principle. However, there were also

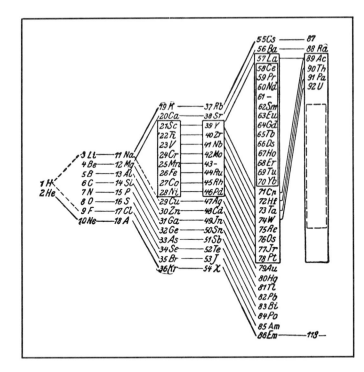

Figure 7.14 The periodic system of the elements according to Bohr (1921). Although this preceded the discovery of Pauli's principle, the figure is essentially correct. [From *Annalen der Physik 71*, 228 (1923).]

problems that proved refractory to any theoretical treatment even though they seemed simple in appearance. An outstanding example of such a problem was the helium atom.

The explanation of the periodic system of the elements was one of the major successes of these times. Bohr had always striven to explain the periodic system by assigning orbits to the atomic electrons, but the criteria used were far from simple and were unreliable. From his earliest papers he kept the problem in mind and attempted various solutions (Figure 7.14). By 1922 he had intuitively assigned orbits that were essentially correct, but he had used arguments that, with hindsight, are not very strong. In particular he accounted for the number and position in the periodic table of the rare earths. This was a difficult problem, and he was proud of the accomplishment. In his Nobel lecture of 1922 he said, "Indeed it is scarcely an exaggeration to say that if the existence of the rare earths had not been established by direct experimental investigation, the occurrence of a family of elements of this character within the sixth period of the natural system of the elements might have been theoretically predicted" [Les Prix Nobel en 1922]. There followed a dramatic confirmation of the correctness of Bohr's orbit assignments. According to him, element 72 should not have been a rare earth, but rather a metal similar to zirconium. There had been several reports of discoveries of a rare earth of atomic number 72 and it had even been named *celtium* by Urbain, although Moseley had not found it in his samples. Here was an opportunity to confirm or disprove Bohr's ideas. It proved unexpectedly easy to find element 72 in zirconium ores; it had the properties of a homolog of zirconium and not of a rare earth. This beautiful piece of work was done in Bohr's institute by the x-ray spectroscopist Coster and by Bohr's old friend Hevesy.

E. C. Stoner, in England, improved on Bohr's work, but nevertheless the assignment of orbits was not made on clear-cut criteria. At this point Wolfgang Pauli (1900–1958) entered the scene. He had made a profound study of the anomalous Zeeman effect and the quantum number assigned to the different levels. He examined the empirical material as classified by Stoner, and at the beginning of 1925 he formulated a new fundamental principle in quantum theory: Two electrons with identical quantum numbers cannot coexist in an atom. Each orbit may be characterized by certain quantum numbers (in the case of atoms these are four in number), and each orbit can either be empty or contain only one electron. Pauli found this so called exclusion principle before the discovery of the spin by studying spectroscopic material. The spin permitted the interpretation of the four quantum numbers in a more natural way than Pauli had surmised. The exclusion principle, or Pauli's principle, as it is often appropriately called, extends much beyond its atomic applications. It provided the key to the explanation of paramagnetism, the behavior of electrons in metals, and many low-temperature phenomena. Later it was also applied to nuclear physics.

In 1926 Enrico Fermi (1901–1954; Figure 7.15) made one of the most fruitful applications of Pauli's principle by incorporating it in statistical mechanics. He thus obtained a new type of statistics, which, like that of Bose and Einstein, applies in nature. Some particles, electrons, for instance, obey Fermi statistics; others, such as alpha particles, obey Bose-Einstein statistics. Many years later, in 1940, Pauli gave a justification of an empirical relationship that had been previously established. Particles with integral spin obey Bose-Einstein statistics (and are called *bosons*), whereas particles with semiintegral spin obey Fermi statistics (and are called *fermions*). Pauli's proof of this fact shows that it has very deep roots in relativity and causality.

In 1926 Dirac gave the quantum mechanical formulation of Pauli's principle and, in the same paper, put Fermi's statistics on a quantum mechanical basis. The previous papers by Pauli and by Fermi had been on semiclassical lines modified by Bohr's principles.

I have mentioned only a few of the applications of atomic models. The flourishing art was reflected in the treatises of the time, first of all in Sommerfeld's *Atombau und Spektrallinien* and subsequently in the many books that derived from it.

Figure 7.15 Enrico Fermi about 1928. (Photo by G. C. Trabacchi.)

The thriving intellectual activity that ultimately led to quantum mechanics centered in Germany. The leading journal was probably *Zeitschrift für Physik,* which was founded after the war. In Germany two schools were particularly important in theoretical physics; one in Munich led by Sommerfeld, and one in Göttingen led by Max Born. The treatise *Atombau und Spektrallinien* was the text everybody studied. One edition followed another, and it was always kept up to date. The work was admirably clear, easy to read, and relatively complete. In those times a good student, after having digested Sommerfeld's book, could go directly to the current literature and begin original research.

A third center was in Copenhagen with Bohr. The best German students often made the pilgrimage from Munich to Göttingen, and finally to Copenhagen. Naturally these three centers of theoretical work were not the only ones. Others existed elsewhere in Germany, in England, in France, in Holland, and in the Scandinavian countries. After 1927 Rome started to gain importance when Fermi was called to a chair there.

Professors from all the leading institutions met frequently and exchanged information not only on technical matters but also on promising new recruits. In the United States the Rockefeller Foundation financed many young scientists, choosing them with uncanny and admirable shrewdness. Among them were several of the founders of quantum mechanics, such as Heisenberg and Pauli. In meetings, in sojourns financed by fellowships, and in short visits, the physicists often established long-lasting friendships. Among the older professors and leaders of schools, Sommerfeld, Ehrenfest, and Bohr distinguished themselves for the interest they took in their pupils and in the talented scholars they met. Not only did they obtain fellowships for them but they also recommended them to their colleagues and to other universities. I still remember the kindness shown to me by Zeeman, who let me use the facilities of his laboratory and subsequently helped me to obtain a fellowship, and I remember the pride and joy I felt when a new edition of *Atombau* reported favorably on some work of mine. For a beginner this was an invaluable boost. Rutherford provided similar sources of encouragment but tended to limit his territory to the British Empire, which was certainly no small domain.

The total number of physicists in the world was relatively small, about one-tenth the present number. The United States appeared on the horizon, especially in the area of experimental work. As already mentioned, R. A. Millikan and A. H. Compton had both performed fundamental experiments, the former by his measurement of the charge of the electron and of Planck's constant, the latter by a spectacular confirmation of the existence of light quanta and their elastic collision with free electrons. American theoreticians were few. They went to study at Göttingen, Copenhagen, and especially at Cambridge, where there was no language barrier. J. R. Oppenheimer, J. C. Slater, and E. U. Condon, among others, came to Europe and later transplanted modern theory to the United States. Some students also came from Japan; an important one was Y. Nishina, who later was most influential in his own country. From the Soviet Union came A. Joffe, J. Fraenkel, and D. Landau, all three future scientific leaders in their country.

Physicists wrote each other letters just as much as they now use the telephone. Some of the correspondence has been published, and much more has been preserved at various places such as the American Institute of Physics and the Bohr Institute. The published correspondence between Born and Einstein, as well as that between Einstein and Sommerfeld, offers vivid testimony to the events.

Bohr's activity continued for a long time after the early models tentatively incorporating quantum theory had been superseded. He played an important role in the development of quantum mechanics, not only through his personal research but also through the inquiring spirit and critical approach he cultivated in his institute and the hospitality he offered many of the most creative young theoreticians from all countries (Figure 7.16). Furthermore, his visits to other centers greatly stimulated scientists and contributed to the diffusion of the new ideas. His first trip to the United States in 1923 is an example of one such visit that brought a living representative of quantum theory to the New World.

Figure 7.16 Enrico Fermi (left) and Niels Bohr in a discussion during a walk along the Appian Way in 1931. (Photo by S. Goudsmit.)

With the advent of Nazism Bohr became one of the most active and efficient rescuers of scientists uprooted by the new form of barbarism. The Copenhagen conferences were now also used as opportunities for finding jobs, even if only temporary ones, for "displaced scholars"—in plain language, people fired chiefly, but not solely, by Hitler. Bohr, with his vast international connections, was very active in this task, which was often successful and of great mutual benefit to the victim and the new employer. Rutherford, as we have said, was the chairman of a British committee devoted to the same purpose, and his friendship with Bohr helped the enterprise.

Later Bohr gave original and fruitful ideas to nuclear theory, but these are not comparable in importance to his earlier youthful papers that revolutionized physics. Nevertheless, his intuition in nuclear research led to simple but prolific models, such as that of the compound nucleus and to its application to nuclear fission. His viewpoints have exerted their influence on the next generation of nuclear physicists, including his son Aage, who now presides with distinction over the Niels Bohr Institute (the old Bohr Institute) and in various ways preserves its traditions.

During the Second World War Bohr had dramatic adventures. Notified by friends that he was about to be arrested by the Nazi occupants of Denmark, he fled the country at night in an open fishing boat, eventually landing in Sweden. Shortly thereafter he was spirited to England in a small military plane that flew very high. In the flight he lost consciousness because of some maladjustment of the oxygen mask, and at one point the crew feared he was dead. From England he moved to the United States and one day, to our surprise, we saw him arrive in Los Alamos at the laboratory devoted to the building of the atomic bomb. Several of the Europeans there were invited in great secrecy to the home of J. R. Oppenheimer (the Laboratory Director) to meet a Mr. Baker; he turned out to be Bohr (and his son, Aage), traveling under false names. There, in a somber atmosphere, we heard first-hand information on occupied Europe. It was the first time we had had such immediate and authoritative testimony about the deeds of the Nazi occupation.

Bohr was deeply concerned about the consequences of atomic weapons, and he tried to influence President Roosevelt and Churchill to reveal the atomic project to the Soviets at an early date, hoping that this gesture would smooth the way to some kind of agreement with the Soviet Union. He wanted to alleviate the possibility of an armaments race and other dangers that he saw looming ahead. At the same time, he hoped nuclear energy could be turned to the benefit of humanity. He had no success. His thoughts on the subject were later expressed in an open letter to the United Nations in 1950.

In his old age Bohr's interests had widened beyond physics. He lived with his wife in Copenhagen in their splendid residence, and the hospitality, courtliness, and the company one met there were reminiscent of the court of a Renaissance prince.

Bohr died suddenly on November 18, 1962; he was 78 years old.

Chapter 8

A True Quantum Mechanics
at Last

The preceding chapter dealt with the triumphs and the tribulations of atomic models. Perhaps nobody was more aware than Niels Bohr of the weaknesses of the methods that had been so successful in unraveling some of the atomic mysteries. I qualify the statement with a "perhaps" because Einstein, too, had original and profound ideas on the subject, on which he had brooded for twenty years without succeeding in finding a lead. By the early 1920s the old methods had reached their limits and a new generation and new forces were required to solve the problem of a consistent quantum mechanics in physics. This was the greatest challenge of the century and required for its solution new ways of thinking. And here we observe a strange occurrence. Within very few years the mystery was attacked from three sides, and at first it seemed that there were three, not one, different yet consistent forms of quantum mechanics. Only a little later was it recognized that these were different mathematical formulations of the same theory and that in fact they were equivalent.

Once the formal mathematical problem had been solved, a new epistemological problem opened up, because the interpretation of the mathematical formulae required comprehensive and deep revision of the basic concept of causality and determinism in physics. It also required a new philosophical analysis no less revolutionary and shattering than that produced by Einstein in his 1905 analysis of space and time.

Louis de Broglie: Matter Waves

The first to take a revolutionary step was a French aristocrat, Prince Louis de Broglie (b. 1892), who was almost unknown among physicists at the time (Figure 8.1). The de Broglie family, originally from Chieri in Piedmont, Italy, has been prominent in French history since the eighteenth century,

Figure 8.1 Louis de Broglie. With his hypothesis on the wave nature of the electron and his famous formula $\lambda = h/p$, de Broglie opened the way to wave mechanics.

producing several marshals, ambassadors, ministers, and, not least, the Duke Maurice de Broglie, the elder brother of Prince Louis. Maurice was a distinguished physicist and the author of early classical studies on x-rays, which he carried out in his palatial home in Paris on rue Byron, part of which he had adapted as a laboratory. In this he reminds me of Lord Rayleigh, his friend, who had adapted his country estate, Terling Place, for a laboratory.

The brothers' parents died when Louis was still very young, and Maurice, older by seventeen years, took paternal care of his brother. Louis's first interests were historical, and he trained seriously in this discipline. However, he gradually lost interest in archival studies. In 1911 he heard his brother, who had been the secretary of the first Solvay Council, comment on the problems of the nature of light, radiation, and quanta, and Louis's scientific interest was aroused. During the First World War he was assigned to radio work in the French army. At the end of the war he started to study physics and leaned toward theory.

De Broglie began to meditate on the dilemma presented by the double nature of light. According to all experiments on interference and diffraction, light consisted of electromagnetic waves, but, according to Einstein's hypothesis, it was corpuscular in all energy exchanges with matter. Both points of view were supported by massive experimental evidence, all earlier results indicating waves and newer results indicating quanta. Some of the most recent data had come from his brother's laboratory and from experiments in which Louis had helped. How could one reconcile these two aspects that appeared so contradictory? As we saw at p. 89, Einstein had already pointed to this as one of the major problems confronting physics.

Louis de Broglie would have liked to participate in the third Solvay Council of 1921, but he did not succeed in wangling an invitation as a guest. Aroused by this refusal, he vowed that by the next conference he would be invited as a participating member because of his discoveries. Indeed, this happened at the fifth meeting in 1927.

De Broglie started his revolutionary deliberations by considering the paradox:

> On the one hand the quantum theory of light cannot be considered satisfactory since it defines the energy of a light corpuscle by the equation $W = h\nu$ containing the frequency ν. Now a purely corpuscular theory contains nothing that enables us to define a frequency; for this reason alone, therefore, we are compelled, in the case of light, to introduce the idea of a corpuscle and that of periodicity simultaneously. On the other hand, determination of the stable motion of electrons in the atom introduces integers; and up to this point the only phenomena involving integers in physics were those of interference and of normal modes of vibration. This fact suggested to me the idea that electrons too could not be considered simply as corpuscles, but that periodicity must be assigned to them also. [Les Prix Nobel en 1929]

He then used relativity to refine his arguments and arrived at the fundamental relation

$$\lambda = \frac{h}{p}$$

that connects the momentum $p = mv$ of a particle to the wavelength λ of a wave that is associated with it.

De Broglie noted further that optics has two faces: There is a geometrical optics that has great formal analogies with classical point mechanics, and there is a wave optics that stresses the undulatory character of light waves. However, one can show that geometrical optics can be derived from wave optics as an approximation. Geometrical optics is commonly used by instrument builders when they trace "rays" of light, and it is valid when the lengths involved are large compared with the wavelength of light. When we analyze the behavior of a lens, we speak of the rays of light and not of the lines perpendicular to the front of the waves.

For over a century it has been well known that there are close mathematical analogies between the rays of light and the trajectories of particles. Around 1835 the Irish mathematician William R. Hamilton wrote the equation of motion of a material point in a field of force in a form very similar to the equations of a light ray in a medium of varying refractive index. In fact, giving different meanings to the symbols used, the equations are the same. A variation of the index of refraction bends the rays of light in the same way as the variation of the potential bends the trajectories of material points.

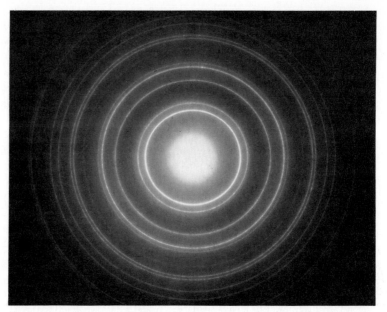

Figure 8.2 Electron diffraction for 100 keV electrons. Their wavelength, given by de Broglie's formula, is about 3.7×10^{-10} cm. (Photo by A. Chambers.)

How is a mechanics to be constructed that bears the same relation to ordinary mechanics that wave optics bears to geometrical optics? This was the problem that confronted de Broglie. From this point of view, he again found the relation $\lambda = h/mv$ mentioned before. Another important result he obtained is based on the idea that if one wants to establish a system of stationary waves on an orbit, the orbit must contain an integral number of wavelengths. Here integral numbers appear, and thus quantization rules arise that, in some cases, give the same results as the Bohr-Sommerfeld quantization rules. The relation between wavelength and momentum is true in general; the quantization method is valid only in special cases and needs further improvements.

Interference phenomena produced by electron waves (Figure 8.2) were first observed, but not recognized as such, by C. J. Davisson and C. H. Kunsman at the Bell Laboratories in New York as early as 1921–1923. After reading de Broglie's papers, Walter Elsasser, then a beginning physicist, explained these experiments as electron diffraction in 1925. When he mentioned his result to Einstein, Einstein answered, "Young man, you are sitting on a gold mine" [W. Elsasser, *Memoirs of a Physicist,* p. 66 (New York: Science History Publications, 1978)]. Clear-cut experimental proof, however, was obtained only in 1927 by Davisson and L. H. Germer at Bell Laboratories and by G. P. Thomson, the son of J. J. Thomson mentioned by

Rutherford (see page 50). Thus both father and son became famous for discoveries pertaining to the electron: the father for its corpuscular aspect, the son for its wave aspect. Discussions in the family were lively and interesting as one may read in the life of J. J. written by his son.

The second part of de Broglie's discovery—his method of quantization—was primitive, but it helped stimulate Erwin Schrödinger in the discovery of the real wave equation of the electron.

The Prince developed his ideas in various notes published in the *Comptes-rendus* of the Académie des Sciences in Paris between 1923 and 1924. He subsequently assembled them in a more complete paper that was intended as his doctoral dissertation. The Sorbonne faculty was embarrassed, as they did not know how to evaluate this thesis. Part of their verdict was: "We praise him for having pursued with remarkable ability an effort which had to be attempted in order to overcome the difficulties besetting the physicists." P. Langevin, the friend of Marie Curie and of Einstein, was impressed, and sent a copy of de Broglie's papers to Einstein. He received a very positive answer; according to Einstein, the paper contained most important discoveries.

We must now temporarily leave de Broglie; later we will see what happened to his approach. In the meantime, we turn to the story of another physicist, about ten years younger than de Broglie, Werner Heisenberg.

Werner Heisenberg and Wolfgang Pauli: Magic Matrices

Werner Heisenberg (1901–1976) was born in Würzburg on December 5, 1901, the son of a professor of Greek at the University of Munich (Figure 8.3). He studied in Munich and at the University became a pupil of Arnold Sommerfeld, under whom he wrote a doctoral dissertation on hydrodynamics. However, he had turned to investigations in atomic physics even before he obtained his degree and had tried to find empirical regularities in spectra, a study that made him master the numerical material available. In Munich, Heisenberg took his recreation in sports that brought him in contact with nature, chiefly skiing and mountaineering. In many ways he was a romantic German patriot with a touch of the Boy Scout spirit and of the German *Wandervogel*. Heisenberg has written an autobiographical book, in which he describes important moments in his life as they appeared to him in his mature years. It presents many echoes of his love of nature and memories of mountain experiences.

Heisenberg not only initiated quantum mechanics but for many years also supplied new and brilliant ideas to many branches of physics. Before he reached the age of thirty he had founded a physics school at Leipzig, where he had been appointed to the chair of theoretical physics; several of the best theoreticians, only a few years younger than he, were educated at his school.

Figure 8.3 Werner Heisenberg (1901–1976) around 1924. His discovery of matrix mechanics and of the relationship between a coordinate q and its conjugate momentum p, $pq - qp = h/2\pi i$ revealed the first complete form of quantum mechanics. (University of Hamburg.)

When the Nazis came to power, Heisenberg, who was in no personal danger, decided to stay in Germany, to which he felt a deep loyalty. The hope of rescuing whatever he could of German science was one of his motives, but he estimated the situation too optimistically and found himself in difficult conflicts. During the war he was one of the leaders of the German atomic bomb project, with indifferent success. After the war he tried hard to help rebuild German science. He moved to Munich, where he directed a large physics research institute, but as has happened to other great physicists, physics had evolved in directions that were not accessible to him. He died in 1976.

When still a student at Munich, Heisenberg met and became a close friend of Wolfgang Pauli, whom we have already mentioned. Pauli was the same age as Heisenberg. He became famous at age twenty-one by writing an article on relativity for the Encyklopädie der mathematischen Wissenschaften (vol. V 19) [*Encyclopedia of Mathematics*]. He was supposed to help Sommerfeld in the task, but the master soon recognized the quality of Pauli's scholarship and left him the job. Einstein was impressed by the article and was surprised when he heard the age of the author. The article has been translated into English, republished in 1958, and is still one of the best texts on relativity.

Pauli was fat, disliked sports, and in many ways was almost the opposite of Heisenberg. Furthermore, he had a scathing critical attitude and a formidable cultural background. He thus served as an oracle and as a judge for checking ideas and results of those who had discovered or believed they had discovered new physics. The oracle, however, was not infallible, as we have seen in connection with the discovery of electron spin by Uhlenbeck and Goudsmit. When, in 1932, some young physicists assembled around Bohr decided to produce a skit adapting Goethe's Faust to the circumstances of modern theoretical physics, Pauli received the role of Mephistopheles, "der Geist der stets verneint" [the spirit that ever denies].

After a period in Göttingen and a stay in Copenhagen in 1923, Pauli took a position in Hamburg, where he established a lifelong friendship with Otto Stern. Later Pauli succeeded Schrödinger in Zurich and, except for the war years spent at the Institute for Advanced Study in Princeton, he remained at Zurich until his untimely death in 1958. He was one of the major theoretical physicists of this century. Among his discoveries were the exclusion principle, the hypothesis of the nuclear spin, the hypothesis of the neutrino, and the unraveling of the connection between spin and statistics of a particle. As a scientific writer, he published, in addition to the treatise on relativity mentioned above, one of the best expositions on quantum mechanics.

Pauli had peculiar manners. For instance, after listening to a lecture at an international conference in which I described some work performed on proton-proton scattering, Pauli left the lecture hall with me and another physicist. He turned to me and said: "I never heard a worse speech than yours." He thought for a moment and then he turned to our colleague and added, "Except when I listened to your inaugural lecture at Zurich." Knowing Pauli well, I was not too worried by his remark. The rocking motion of his body during my lecture was evidence that he had listened, and this meant more than his comments, as anybody familiar with him knew. When one presented a new idea or theory to Pauli, he almost automatically answered, "Rubbish" (*Quatsch,* in German). However, if one insisted on the idea and it became possible to convince him of its validity, then he became helpful and offered new points of view. He had many friends who had learned how to deal with him, and their high opinion of him made up for his initial gruff behavior.

Pauli and Heisenberg corresponded continuously, when they could not speak directly to each other, and their letters are a dramatic commentary to the development of quantum mechanics. Pauli ranges from enthusiastic to sharply critical of his friend. Pauli also had an abiding interest in psychology, corresponded with Carl Gustav Jung, and even wrote articles on psychological subjects. This other aspect of his personality, although apparently very important to him, was kept rather private.

In 1922, when he was still Sommerfeld's student, Heisenberg accompanied his mentor to Göttingen to listen to some lectures by Bohr. After one of these lectures Heisenberg had a long discussion with Bohr, a discussion that was further prolonged during a hike they took in the vicinity of Göttingen. Bohr was impressed by the young scholar and invited him to

Copenhagen. Heisenberg obtained his doctorate degree at Munich and immediately thereafter went to Göttingen to study further with Born. He remained at Göttingen until the fall of 1924, except for a visit to Copenhagen at Easter in the same year.

In 1924, following a study of light-dispersion theory undertaken with Bohr and Kramers, Heisenberg became suspicious of several intuitive concepts used in quantum theory, such as the literal picture of electronic orbits in atoms. Pauli, in his correspondence with Heisenberg, expressed similar misgivings. Heisenberg then strove to formulate a theory avoiding the concrete but unobservable representations of the orbits, using only observable quantities, such as the transition probabilities for quantum jumps. In this way he hoped to come closer to reality and avoid spurious concepts based on models. This formulation brought about the need to use quantities labeled by two indexes, corresponding to initial and final state of the system considered; these quantities were also connected to the Fourier expansion of the coordinates representing the periodic motion in the orbit model. The scheme generated a type of algebra that, to Heisenberg's surprise, was not commutative; that is, the product of two quantities depended on their order of multiplication. Heisenberg wrote a paper developing these ideas in May 1925, during a stay on the island of Helgoland in the North Sea, where he had repaired to escape from pollens that gave him a painful allergy.

At that time Heisenberg did not know the mathematical theory of matrices, but when he spoke about his noncommuting quantities to Max Born (1882–1970), the latter soon recognized that they had to do with matrix algebra, well known to Born (Figure 8.4) from his student days. Heisenberg, Born, and Pascual Jordan, another mathematically gifted student of Born, joined forces and soon obtained a consistent scheme of quantum mechanics that gave correct results. Their method was based on a refinement and deeper interpretation of the correspondence principle joined to the use of matrices for the representation of kinematic variables. The method was also connected to Hamilton's classical analytical mechanics. The coordinates q and their conjugate momenta p, however, were not ordinary numbers varying with time, but matrices—that is, arrays of complex numbers disposed in a square. The position of each number in the square is denoted by two indexes, the first corresponding to the row and the second to the column in which the number is located. In the specific physical application the indexes correspond to initial and final state of a system. Mathematical objects such as matrices can be added and subtracted according to simple rules completely similar to those prevailing for ordinary numbers. The multiplication rule, however, is peculiar. Multiplication of the matrix p by the matrix q does not give the same result as the multiplication of q by p. On the contrary, the matrices obey the strange commutation rule:

$$pq - qp = \frac{h}{2\pi i}$$

Figure 8.4 Max Born (1882–1970) at Lake Como in 1927 during an excursion organized for the participants in the Volta Conference. He had just provided the probabilistic interpretation of wave mechanics. (Courtesy of F. Rasetti.)

In greater detail, for the matrix elements p_{mr}, q_{mr}, this equation is written as

$$\sum_r (p_{mr}q_{rn} - q_{mr}p_{rn}) = \frac{h}{2\pi i} \delta(m,n)$$

where $\delta(m,n)$ is a Kronecker symbol, equal to 1 if $n = m$ but otherwise equal to zero. In terms of classical physics for a point moving in one dimension, q is the coordinate x, p is the momentum $m\dot{x} = mv$. In classical physics when p and q are numbers, obviously $pq - qp = 0$. However in quantum mechanics p and q are more complicated expressions—matrices—and they obey the commutation formula given above.

This new mechanics, called *Matrizenmechanik* or *Quantenmechanik*, was far from clear and presented great difficulties in computation. Besides having studied the general theory, Heisenberg and his colleagues had applied it, as specific examples, to the harmonic and anharmonic oscillator and to some other simple problems. With a tremendous effort Pauli had succeeded in applying it to the hydrogen atom, achieving by the new method the results that Bohr had obtained in 1912 by using inconsistent hypotheses.

Quantenmechanik contained fundamental new ideas that had a strong appeal. It seemed to open new vistas, especially as it offered the possibility of abandoning the orbit concept and of using only observables, but its physical concepts were somewhat nebulous and it did not solve new concrete problems, at least at the beginning. All told, its appeal was limited to a relatively small group of initiates.

Figure 8.5 Paul A. M. Dirac
(b. 1902) in 1934. Dirac gave
general formulation of quantum
mechanics, and his relativistic
equation for the electron had
profound and long-lasting
consequences. (Photo Ramsey &
Muspratt.)

Paul Adrien Maurice Dirac: Abstraction and Mathematical Beauty

While these events were taking place between Göttingen and Copenhagen, a surprise was being prepared in Cambridge. A new and, at first sight, different quantum mechanics was developing. Its discoverer was once more a young man, little known to physicists: Paul A. M. Dirac (Figure 8.5). Born August 8, 1902, at Bristol, England, from a Swiss father and a British mother, Dirac was a contemporary of Heisenberg, Pauli, and Fermi. Dirac had begun to study electrical engineering at Bristol, but he changed to pure mathematics while he was still there and later pursued mathematics at St. John's College, Cambridge, where he became an 1851 Exhibition Senior Research Student. At Cambridge he had learned Bohr's atomic theory and written some papers on the subject. In 1925, after Heisenberg visited Cambridge, Dirac received the proofs of Heisenberg's first paper on *Matrizen-mechanik,* which was Dirac's first introduction to quantum mechanics. After studying the proofs for about ten days, he came to the conclusion that noncommutation was the essential new idea. Here are Dirac's own words:

> For some time I puzzled over this very general relationship to try to see how to make a connection with the laws of mechanics which were already well understood. At this time I used to take long walks on Sundays alone, thinking about

these problems and it was during one such walk that the idea occurred to me that the commutator A times B minus B times A was very similar to the Poisson bracket which one has in classical mechanics when one formulates the equations in the Hamiltonian form. That was an idea that I just jumped at as soon as it occurred to me. But then I was held back by the fact that I did not know very well what was a Poisson bracket. It was something which I had read about in advanced books of dynamics, but there was not really very much use for it, and after reading about it, it had slipped out of my mind and I did not very well remember what the situation was. It became necessary to check whether the Poisson bracket really could be made to correspond to the commutator and I needed to have a precise definition of the Poisson bracket.

Well, I hurried home and looked through all my books and papers and could not find any reference in them to Poisson brackets. The books that I had were all too elementary. It was a Sunday, I could not go to a library then; I just had to wait impatiently through that night and then the next morning early, when the libraries opened, I went and checked what a Poisson bracket really is and found that it was as I had thought and that one could set up the connection between a Poisson bracket and a commutator. This provided a very close connection between the ordinary classical mechanics which people were used to and the new mechanics involving the noncommuting quantities which had been introduced by Heisenberg.

After this early idea, the work was all fairly straightforward. There were really no serious difficulties for quite a long time. One could work out the equations of the new mechanics; one just had to make the appropriate generalization in the classical equations expressed in the Hamiltonian form. I continued developing this work and Heisenberg and the people working with him developed the matrix point of view in Göttingen independently; we had some correspondence but we were working essentially independently. [*Proceedings of the International School of Physics,* "Enrico Fermi," vol. 57, p. 134]

There are certain numbers that are essential to Dirac's form of quantum mechanics. He called them q-numbers in order to distinguish them from ordinary or c-numbers; q-numbers have a noncommutative algebra and are strictly connected to Heisenberg's matrices and to Schrödinger's operators, which we shall mention later. The letter q stands for quantum; c stands for classical.

Thus as early as 1925 Dirac succeeded in giving a complete formulation of quantum mechanics that in many ways was more general than that of his contemporaries. It is remarkable for its axiomatic formulation and for the generalizations it permits.

In 1932 Dirac became Lucasian Professor of Mathematics at Cambridge University, the same chair occupied by Newton in the eighteenth century, and he remained there until his retirement. A rather silent character, he exerted a profound influence through his research papers and through his book *The Principles of Quantum Mechanics*. His writings are characterized by a concise and profound style that often requires concentrated effort from the reader. An anecdote, which possibly is authentic, serves to describe the man. At the end of a seminar Dirac asked, as usual, if there were any questions. Somebody from the audience ventured, "I have not understood how you passed from A to B," indicating two equations. Dirac coolly answered "That is a statement, not a question."

Erwin Schrödinger

Unbeknown to the Göttingen, Copenhagen, and Cambridge physicists, in Zurich another physicist named Erwin Schrödinger (1887–1961) was also discovering quantum mechanics, albeit under a different form. Schrödinger (Figure 8.6), born in Vienna, was already a well-known physicist with distinguished results to his credit. He had an artistic temperament and wrote easily and clearly. His Austrian father was a cultivated gentleman, well versed in many scientific subjects, including chemistry and botany; his mother was English. Erwin had studied in Vienna, where the memory and spirit of Boltzmann were still alive. Schrödinger's teacher had been Fritz Hasenoehrl, himself Boltzmann's pupil and a physicist of great promise. Hasenoehrl unfortunately was killed in the First World War. Schrödinger had fought in the Austrian army in that war. Later he had worked at several universities and finally settled at the University of Zurich, where he carried out his immortal work. Impressed by de Broglie's ideas and by some favorable comments of Debye and of Einstein pertaining to them, he developed them into a true wave theory. At first, he tried to treat everything relativistically, but he found that the results disagreed with experimental evidence; he grew discouraged and temporarily abandoned his efforts.

After some months, Schrödinger returned to the subject using a nonrelativistic approximation and found results that agreed with experimental data. The reason his relativistic theory did not agree was that he did not take into account the electron spin. The idea of the spin was quite new at that time, and theoreticians were not accustomed to it.

He published his work in January 1926 [*Annalen der Physik 79,* 361 (1926)] and wrote the famous Schrödinger equation:

$$\nabla^2 \psi(x,y,z) + \frac{8\pi^2 m}{h^2} [E - U(x,y,z)]\psi(x,y,z) = 0$$

This equation has a form that was well known to earlier mathematical physicists who had studied waves. In fact, it is the typical equation of all wave motions: Acoustic waves, electromagnetic waves, and so on are all treated by equations mathematically very similar to Schrödinger's. Furthermore, just before the appearance of Schrödinger's papers, Richard Courant and David Hilbert, two famous mathematicians from Göttingen, published a book entitled *Methoden der Mathematischen Physik,* which contained all the mathematical underpinnings necessary for an understanding of Schrödinger's papers. In fact, Schrödinger himself found the mathematical solution to problems deriving from special cases of his equations in Courant and Hilbert's book.

Figure 8.6 Erwin Schrödinger (1887–1961), who formulated the wave version of quantum mechanics and found the equation bearing his name. This was a versatile and rich instrument, which explained many phenomena. (Courtesy of W. L. Scott.)

In Schrödinger's equation ∇^2 is the Laplace operator $\partial^2/\partial x^2 + \partial^2/\partial y^2 + \partial^2/\partial z^2$, E is the energy of the system, and $U(x,y,z)$ is the potential that characterizes the system. For instance, for a hydrogen atom $U(x,y,z,) = Ze^2/r = Ze^2/\sqrt{(x^2 + y^2 + z^2)}$. The function $\psi(x,y,z)$ was called the *field scalar* by Schrödinger. Its interpretation will be discussed later. Schrödinger's equation has "acceptable solutions"; that is, solutions such that $\int |\psi(x,y,z)|^2\, dx\, dy\, dz$ is finite only for certain special values of E. These values are called *eigenvalues,* and their ensemble forms a spectrum. In the case of hydrogen there is a discrete spectrum identical to Bohr's energy levels and a continuum for positive energy. The quantization thus happens automatically. The energy levels become similar to the proper frequencies of a vibrating string, that is, to the tunes emitted by any string instrument.

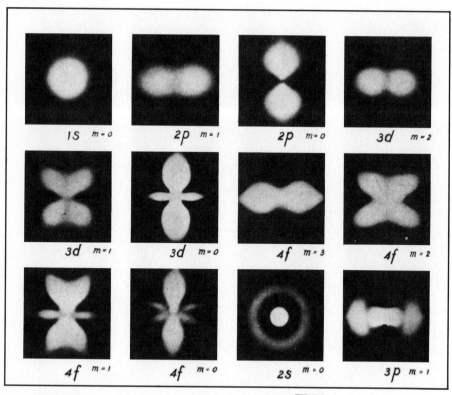

Figure 8.7 Schrödinger's electron clouds, corresponding to stationary states of a hydrogen atom. They indicate the "electron density," that is, the probability $|\psi|^2$ of finding the electron at a certain point when the atom is in the states given by the specified quantum numbers. The first number is n, the total quantum number; the letters $s, p, d, f,$ for historical reasons, indicate the value of the angular momentum in units of $h/2\pi$; s corresponds to $l = 0, p$ to $l = 1, d$ to $l = 2, f$ to $l = 3$. m is the magnetic quantum number that gives the orientation of the atom. [From H. White, *Physical Review, 37,* 1416 (1937).]

Schrödinger's papers appeared as a series of memoirs in the *Annalen der Physik,* the same journal that had published Planck's and Einstein's memorable discoveries. Their collective title is "Quantization as an Eigenvalue Problem," and they immediately attracted universal attention and admiration. The letters between Schrödinger, Einstein, Planck, and other luminaries bear witness to this. Planck invited Schrödinger to Berlin to address his seminar, and it was moving to see Planck, already at an advanced age, rejoice in finally seeing a definitive and consistent form of quantum mechanics.

Why was Schrödinger's success so immediate and universal compared with the more modest acceptance of Heisenberg's earlier work? One

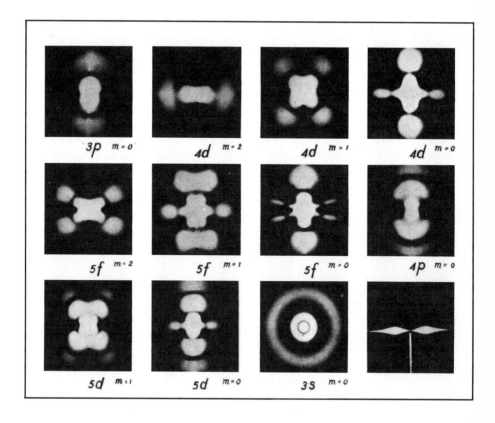

reason was certainly that Schrödinger's mathematics was of a type familiar to physicists and that his whole method was not mathematically different from classical wave theory. Even the young Fermi had serious difficulties in grasping Heisenberg's ideas; by contrast, he immediately assimilated Schrödinger's papers, and, as they were published, he explained them to his friends and pupils. Fermi was hindered not by Heisenberg's mathematics, which he easily mastered, but by Heisenberg's physical ideas. Another reason for Schrödinger's success is that his methods could be applied to concrete practical problems much more easily than Heisenberg's methods, and thus they could be compared with experiments.

However, there was a big unsolved question. What is the ψ, the mysterious field scalar that propagates as a wave? For some time Schrödinger and others thought that the square of the modulus of the complex number ψ was the density of electric charge, as if the electron dissolved itself into a cloud. This interpretation, however, was very suspect because there was good reason to believe that electrons are localized in very small regions of space, almost as points. Figure 7.5 shows Bohr's orbits and Figure 8.7 Schrödinger's clouds of electricity for the hydrogen atom, neither of which exist; the only solid concept is the mathematical abstraction embodied by the formulae. The formulae can be translated to apply to experimental situation, but the other concepts cannot.

In Copenhagen Schrödinger's theory was admired, but his interpretation of the ψ was rejected. Bohr invited Schrödinger to discuss problems connected with his theory and confronted him with criticism and objections to the interpretation. The protracted discussions were so laborious that at the end Schrödinger fell ill from exhaustion and went to bed in Bohr's residence, where he was staying. This did not stop his host from continuing the argument with Schrödinger in his bedroom. Remember that Bohr was a kind soul, polite and sincerely concerned for the well-being of his guest, but on such a vital problem of physics he could not control himself.

Schrödinger's later life is a mirror of the times in which he lived. He succeeded Planck in Berlin, but in 1933, disgusted by the advent of Nazism, he left Germany and went to Oxford. Imprudently he returned to Graz in his native Austria in 1936, only to be forced from his country in 1938 when Hitler annexed Austria to Germany. He arrived in Rome with a rucksack that contained the only possessions he had been able to take, and sought out Fermi, requesting him to take him to the Vatican, where he found shelter and a discrete asylum for a short time. From there he went to the Institute for Advanced Study in Dublin, where he remained until 1955. In Dublin he composed a booklet, "What Is Life?" that exerted considerable influence on biophysicists. Later he returned to Vienna and died there in 1961.

Let us return to ψ. Its usefulness and its puzzling nature are well described in a quatrain composed by Walter Hückel a young collaborator of Debye:

> Erwin with his psi can do
> Calculations quite a few.
> But one thing has not been seen
> Just what does psi really mean.
> (Felix Bloch's translation)

The meaning of ψ started to become clear in 1926 when Max Born interpreted it not as electricity density, but as probability density; that is, $|\psi(x,y,z)|^2 d\tau$ is the probability of finding the electron in a volume element $d\tau$ at coordinates x, y, z. Naturally, this idea posed startling problems because it was new that a mechanical theory should give a probability, although in 1917 Einstein with his A and B had foreshadowed the role of probability in atomic phenomena.

A big hurdle was successfully overcome when Schrödinger, the young American Carl Eckart, and others independently recognized that Heisenberg's and Schrödinger's theories were mathematically equivalent. If one knew how to solve Schrödinger's equation for a certain problem, one could calculate Heisenberg's matrices, and vice versa. It was almost like solving the same problems by geometrical or by analytical means. Hence, wave mechanics and *Matrizenmechanik* were equivalent; no, they were the same thing.

Dirac's theory, too, is equivalent to Heisenberg's and Schrödinger's theories. For all three the essential relation that produces the quantization is

$$pq - qp = \frac{h}{2\pi i}$$

For Heisenberg p and q are matrices; for Schrödinger q is a number and p is the differential operator

$$p = \frac{h}{2\pi i} \frac{\partial}{\partial q}$$

For Dirac p and q are special numbers obeying a noncommutative algebra. The results of any calculation on a concrete problem done by any of the three methods are identical.

The Meaning of the Equations

The problem of the interpretation of the mathematics is a tougher one. It is a fact that both light and matter show some characteristics of particles and some characteristics of waves. Here we are referring to classical macroscopic particles and classical macroscopic waves. We can then ask, Is light a wave phenomenon or is light composed of particles? A similar question can be asked for, say, an electron.

To answer such questions, we must resort to a way of thinking and to methods introduced by Einstein in 1905, when he showed how to analyze the seemingly simple concept of simultaneity and arrived at the consequent startling results.

Using similar methods, as Heisenberg first showed in a famous paper of 1927, we find that classical concepts such as that of the orbit of a particle or that of classical wave concepts fail when we try to apply them to microscopic objects. By doing certain experiments, we may "show" that we are dealing with a wave, but then it becomes impossible to perform on the same object the experiments necessary to show that we are dealing with a corpuscle, and vice versa. The quantum conditions are at the root of this dilemma.

If we try simultaneously to determine the coordinate and the momentum of a particle, we encounter contradictions originating from the double nature—corpuscular and wave—of the particle. In order to illustrate this difficulty Heisenberg discussed many experiments that show that this measurement is impossible to achieve because of the double nature (wave and corpuscle) of a particle such as an electron. The precision of the result is limited by a famous relation called the uncertainty principle, established by

Heisenberg. Calling Δq the inescapable error affecting the coordinate measurement and Δp the analogous error for the momentum measurement, we have

$$\Delta p \; \Delta q \simeq \frac{h}{2\pi}$$

The term *error,* of course, does not mean the error caused by the imperfection of practical instruments, but the error inherent in inescapable phenomena, such as the recoil on emission of a quantum or the diffraction of a wave through a slit.

The uncertainty equation is strictly connected to the quantum condition $pq - qp = h/2\pi i$. Uncertainty relations obtain only for magnitudes whose matrices do not commute. Heisenberg's epochal paper, in which he set down these ideas, was acclaimed by Bohr; Pauli also applauded the theory, which had been communicated to him in a letter.

In our everyday life, where we deal with objects of human dimensions, we have no occasion to experience the limitations imposed by the uncertainty principle, and we are unaware of them. If we observe a macroscopic object, the perturbation in its behavior introduced by observation is completely negligible, and thus we escape the logical strictures imposed by the uncertainty principle. The recoil on emitting a quantum that goes to my eye (and that I need to perceive the object) is negligible if the emitter is even a small piece of a crystal, and thus I do not have to worry about the change of momentum that my observation forces on it. But when the emitting particle is an atomic electron, the recoil is large enough to affect its subsequent motion, and hence the observation perturbs the observed object. In this case a critical examination of the procedures that one can theoretically use in measuring coordinate and momentum lead to the uncertainty relation.

We can understand this best through an example. Let us assume that we have many identical systems or that we repeat the same experiment many times on the same system. We use perfect instruments, in the sense that we do not consider technical limitations. However, we can not ignore limitations imposed by the laws of nature. We measure the coordinate of an electron many times and each time we find the value q; the average value of q is $\langle q \rangle$. We define Δq as $\sqrt{\langle (q - \langle q \rangle)^2 \rangle}$, or the root of the mean square deviation of $q,$ and similarly for p. Now let us assume that our measurement consists of observing that the electron passes through a slit of width a in the x direction; then $\Delta x = a$. Assume the electron is sent through the slit with a velocity in the y direction v_y, perfectly perpendicular to the plane of the slit. After passing through the slit, however, the electron may have a different direction from the original one because the electrons are subject to diffraction. It thus acquires a velocity in the x direction, given by the diffraction width, which is approximately $(\lambda/a)v_y$. For λ we must insert the de Broglie wavelength h/mv_y, and from these relations we immediately obtain

$$\Delta x \; \Delta p_x = am \; \frac{\lambda}{a} \; v_y = h$$

If we narrow the slit, making a small, Δp_x grows just as much as a decreases. The root of the matter is that material points behave both as particles and as waves, and thus it is impossible to give an accurate description from only one point of view.

There is also another uncertainty relation that states that in the time interval Δt one can measure an energy with an accuracy ΔE such that

$$\Delta t \; \Delta E \simeq h$$

In addition to these limitations, unknown to classical physics, and because of them, quantum mechanical problems are formulated in a completely different way from the usual macroscopic formulation. For example, we have a mechanical system such as a hydrogen atom (we will neglect spins, for simplicity). It is futile to ask about orbits or other concepts that are not susceptible to measurement because of the double nature of the electron. However, we may inquire about energy or angular momentum or other quantities that are measurable. Theory gives us a catalog of possible values of the results of a measurement. In the specific case of the energy of a hydrogen atom those are the negative values $-2\pi^2 m z^2 e^4 / h^2 n^2$ and all positive values. (The zero of the energy is when the electron is at rest at great distance from the nucleus.) For angular momentum the catalog gives $lh/2\pi$ where l is an integral number, zero or positive. Furthermore, the theory gives the probability under certain circumstances of finding any given value specified in the catalog. In classical physics, on the other hand, the catalog contains a continuum of values, and we can specify exactly which one will be obtained under certain circumstances.

For a more technical, very condensed summary of the quantum mechanical formulation of physical problems, see Appendix 10.

A New Look at Reality: Complementarity

Bohr further deepened the basic ideas of quantum mechanics, reaching the notion of complementarity. Two magnitudes are complementary when the measurement of one of them prevents the accurate simultaneous measurement of the other. Similarly, two concepts are complementary when one imposes limitations on the other. Just as the correspondence principle was hard to formulate precisely, but rather represented Bohr's way of thinking, in the same fashion complementarity is more a way of thinking than a precise concept.

In 1930 Heisenberg summarized the relations between all these trends in ideas in the table below [Heisenberg, *Physical Principles of the Quantum Theory*].

Classical theory	Quantum theory		
Space-time description Causality	I Alternative Space-time description Uncertainty relation	Statistical connection	II Alternative Abstract mathematical scheme—no space-time description Causality

The new formulation of physical problems typical of quantum mechanics and the type of answers quantum mechanics gives may at first sight appear unsatisfactory. This is mainly because they do not correspond to the usual images we have in our mind, based, as they are, on macroscopic everyday experience. However, the theory adheres strictly to reality defined as what can be experimentally observed. This way of thinking introduces in an essential manner the concept of probability, even if the probability occasionally may be one—in other words, certitude. It requires a rejection of our habits of thought similar to that required by Einstein's 1905 analysis of space-time notions, based on special relativity.

Not everybody likes upheavals of this kind. Even Einstein, who had paved the way, resisted both probabilistic physics and the formulation of quantum mechanics according to Bohr and other physicists of the Copenhagen school. He disliked it and thought that it might be superseded by something more deterministic, although of course he recognized its logical consistency and its agreement with experimental evidence. With the passing of time, however, the new generations adapted to quantum mechanics. Those who preserved a skeptical attitude were, remarkably, some of the founders of the new doctrine; de Broglie, Schrödinger, and to some extent Dirac.

Quantum mechanics underwent an almost official inauguration at the International Physics Conference held at Como (Italy) in 1927 on the hundredth anniversary of Volta's death (Volta came from Como). Many of the most important physicists of the time were present, including several of the younger generation (Figure 8.8). Bohr spoke on quantum mechanics, delving especially into epistemological questions. Conspicuously absent was Einstein, an irreplaceable absence, but Einstein did not want to set foot in Mussolini's domain. A few weeks later, however, Einstein was present at Brussels for the Solvay Council, together with all the authorities on quantum mechanics: Heisenberg, Dirac, Pauli, Bohr, Born, and others (Fig-

Figure 8.8 Left to right: Fermi, Heisenberg, and Pauli at Lake Como in 1927. Heisenberg had recently discovered matrix mechanics and the uncertainty principle. Pauli had just discovered the exclusion principle, and Fermi had just discovered the statistics based on this principle. (Courtesy of F. Rasetti.)

ure 8.9). Einstein tried all possible means to concoct counterexamples to the uncertainty principle, thus attempting to undermine the foundations of the new theory. Every morning at breakfast he would present Bohr with an ingenious example that he thought would offer a contradiction. Bohr studied it until he succeeded in finding fault with Einstein's criticism, only to find that Einstein had prepared another example. One of the most subtle and difficult examples could be solved only with the help of general relativity, Einstein's own creation. In the end Einstein had to admit that he had not found a valid counterexample, but he maintained his belief that "God does not play dice," as he wrote in a private letter to Born.

Heisenberg, Pauli, Dirac, and other founders of quantum mechanics spent considerable time in Bohr's institute at Copenhagen during the decisive period for the development of the theory. Their long discussions with Bohr and among themselves helped clarify the ideas and crystallize the doctrine. The intellectual communion, the recollection of difficulties overcome with mutual help, and the common life generated an esprit de corps and produced a scientific "Copenhagen spirit." This spirit showed a certain propensity to transform itself into an orthodoxy, as often happens under such circumstances. However, this orthodoxy was not absolutely rigid and

Figure 8.9 The participants in the 1927 Solvay Council. It was devoted to quantum mechanics, and the new field was, so to speak, officially inaugurated at this meeting. There were discussions between Einstein and Bohr. In the front row, left to right: I. Langmuir, M. Planck, M. Curie, H. A. Lorentz, A. Einstein, P. Langevin, C. E. Guye, C. T. R. Wilson, O. W. Richardson. Second row, left to right: P. Debye, M. Knudsen, W. L. Bragg, H. A. Kramers, P. A. M. Dirac, A. H. Compton, L. V. de Broglie, M. Born, N. Bohr. Standing, left to right: A. Piccard, E. Henriot, P. Ehrenfest, E. Herzen, T. De Donder, E. Schrödinger, E. Verschaffelt, W. Pauli, W. Heisenberg, R. H. Fowler, L. Brillouin. (Institut Solvay.)

tended, in typical Bohrian fashion, to have several variants. It was not for nothing that Bohr loved soft contours. I would almost say he liked Danish fogs, and I mention one of his favorite quotes:

Nur die Fuelle fuehrt zur Klarheit   [Only fullness brings to clarity
Und im Abgrund liegt die Wahrheit   And the truth lies in the Abyss]

I will add a little story about Bohr, even if it has been told many times, because it illustrates his personality and his type of humor. Bohr had a summer home where he spent his vacations; on a door of this house he had put a horseshoe. One of his visitors, somewhat surprised, asked him if he believed that horseshoes brought good luck. Bohr answered "No, but I was told that they also bring luck to people who do not believe in them."

Until 1927 no form of quantum mechanics was relativistic. Physicists, starting with de Broglie, had tried in vain to find a relativistic theory. All attempts either brought absurd results or departed from experimental evidence.

In 1928 Dirac finally found a mathematical way of writing a relativistically invariant equation for an electron. He found that the mathematics introduced a new internal degree of freedom of the particle. This degree of freedom turns out to have all the properties of the electron spin, starting from its value $h/4\pi$. It also has a magnetic moment of value $eh/4\pi mc$. This almost miraculous result gives great strength to Dirac's theory. In the previous nonrelativistic theories spin and magnetic moment had to be introduced as separate ad hoc pieces of information; here everything came out automatically. But there was more. Dirac's equation describes not only the motion of electrons, but also that of particles with the same mass as the electron but with a positive charge. Such particles had not been seen at the time Dirac formulated his theory. Dirac tried to identify these unwanted particles as protons, but he met with grave difficulties. Pauli, in his *Treatise on Quantum Mechanics,* expressed the opinion that "an attempt to save the theory in its present form appears hopeless in the face of its consequences ..." As we shall see later, positive electrons do in fact exist, and when, in 1932, they were discovered in cosmic rays, the defect in Dirac's theory became one of its triumphs. Dirac's theory requires new mathematical devices called *spinors,* which have some analogies with 4-component vectors.

With these papers by Dirac the main lines of quantum mechanics were established. The result is a description of nature both more abstract than that expressed through classical physics and at the same time much closer to experiment.

We can see a curious parallel between the world of art and the world of science. During Planck's lifetime, the Impressionists had developed a new style of painting. Around the time of the formation of quantum mechanics, artists such as Picasso were tending further toward abstraction, depicting, for example, a figure with two faces. One could hardly say, however, that one face is wavelike and the other corpuscular.

Mysteries Explained, But Doubts Remain

All physics breakthroughs (I dislike this much-abused word but must use it when it is truly appropriate) bring not only general theories, but also the explanation of a host of specific phenomena that were empirically known but could only be treated phenomenologically, often by the introduction of ad hoc hypotheses and empirical coefficients. This blossoming of applications also happened for quantum mechanics. By applications I do not mean technological applications; those came later. I mean the explanation of phenomena that had been known for a long time but had remained unexplained.

The theory of the helium atom was one of the first of Heisenberg's successes. In it we meet several unexpected features for the first time. For instance, in the helium atom there are two families of energy levels and it is not possible to pass from one to the other by the emission of radiation. In one system the spin of the two electrons are parallel to each other; in the other they are antiparallel. The energy difference between the two types of levels was totally unexplainable with the old Bohr-type models. But

Heisenberg simply calculated the wave functions and energies according to the rules of quantum mechanics, without bothering to understand the phenomenon from an orbit point of view, and found the right answer. The energy difference, which is connected to the identity of the electrons, is based on phenomena that do not have a classical interpretation.

This explanation was followed by the theory of homeopolar chemical binding in molecules such as the hydrogen molecule. It was known from Avogadro's time that the hydrogen molecule was diatomic, but why did two atoms of hydrogen attract and bind each other yet a third would not bind with them? Why do hydrogen or oxygen form molecules, but noble gases do not? These are some of the simplest questions of theoretical chemistry, but until 1927 there was no answer to them. Walter Heitler and Fritz London, following Heisenberg's lead, presented the first theory of chemical binding of identical atoms. This theory was developed into a whole branch of science and in time produced spectacular results. Linus Pauling was one of its pioneers on the chemical side.

Many other old problems capitulated under the onslaught of the new quantum mechanics. Collision theory was treated by Born, and it was this study that gave the clue to the fundamental statistical interpretation of the ψ. Dirac gave a profound quantum mechanical interpretation to Pauli's exclusion principle, at the same time rediscovering Fermi's statistics. The theory of paramagnetism was elucidated by Pauli. Felix Bloch made a giant step in the theory of metals by calculating how an electron wave would move in a periodic potential field. Heisenberg explained ferromagnetism, justifying a previous phenomenological theory that arbitrarily introduced in a ferromagnetic substance such as iron a huge internal magnetic field. Last but not least, George Gamow, and Edward U. Condon with R. W. Gurney independently discovered the transparency of potential barriers and applied it to the explanation of the apparent paradoxes of nuclear alpha decay. This result deeply impressed Rutherford, who had been somewhat skeptical about a theory involving so much mathematics and so little intuition.

There was a period of a few years when to be able to make true discoveries it was sufficient to understand quantum mechanics and to know its techniques no better than a good graduate student would today. Of course, at that time it was not so easy to master even the elements of quantum mechanics. After this period of tumultuous development there was a great need for systematization. The original discovery papers are often difficult to read, but toward 1929 books or monographs started to appear in which the various points of view were explained more systematically. Dirac, Heisenberg, Born, Jordan, and Pauli each wrote either a book or a monograph on the subject. These works form the basis of the modern treatises used by today's students. Sommerfeld, too, added a supplementary volume to *Atombau und Spektrallinien*.

Non-relativistic quantum mechanics is by now a closed discipline, at least in principle. It is a grandiose generalization of classical mechanics, on which it is modeled to a considerable extent, and it vindicates the correspondence principle because for macroscopic objects it gives the same answers as classical theory.

Relativistic quantum mechancs is much less advanced. Dirac's theory is limited to particles of spin $\frac{1}{2}$; for other spins there are difficulties. But even for spin $\frac{1}{2}$ particles in a given preassigned electromagnetic field, there are problems. It is possible to develop a perturbation theory, which, however, diverges. We can remedy the situation by introducing a cut-off at high energy, but this spoils the relativistic invariance. Nevertheless, we reach extremely great precision and, for instance, the magnetic moment of the electron or of the muon can be calculated to 0.3 parts per billion. The result agrees with experiments within an error of 0.2 parts per billion. These are the best-known measured and calculated numbers in physics.

However, in 1972 Dirac closed a conference on the development of quantum mechanics with these words:

> Now, what can we make of this situation? It seems to me to be evident that we do not yet have the fundamental laws of quantum mechanics. The laws that we are now using will need to have some important modification made in them before we shall have a relativistic theory. It is very likely that this modification from the present quantum mechanics to the relativistic quantum mechanics of the future will be just as drastic as the modification from the Bohr orbit theory to the present quantum mechanics. When we make such a drastic alteration of course our ideas of the physical interpretation of the theory with its statistical calculations may very well be modified. [Dirac, "The Development of Quantum Mechanics," Acc. Naz. Lincei, Roma, 1974]

Chapter 9

The Wonder Year 1932:
Neutron, Positron, Deuterium,
and Other Discoveries

The study of quantum mechanics from the publication of Planck's paper in 1900 to Bohr's in 1913 had engrossed relatively few physicists. On the other hand, from 1913 to the completion of the theory in 1928, it absorbed the entire strength of the new generation of physicists, with the possible exception of those centered around Rutherford. In this period theory prevailed over experiment, although without experimental support nothing would stand. In order to believe the startling theoretical results, so different from everyday common-sense experience based on macroscopic objects, experimental evidence was absolutely necessary; the apparently paradoxical Stern-Gerlach experiment is a case in point.

After Dirac's relativistic theory of the electron in 1928 there was the feeling that a turning point in physics had been reached. That this feeling really existed at the time and was not imagined afterward is demonstrated, for instance, by an interesting speech of Professor Orso Mario Corbino (1876–1937), then the director of the Physics Institute at the University of Rome. The Italian Society for the Advancement of Science held an annual convention in which it endeavored to inform "intelligent laymen" of what was happening in various sciences. Fermi, E. Persico, and others had repeatedly spoken of the new physics, trying to explain its results to the Italian public. In 1929 Corbino decided to survey more general questions on the future of physics, and after extended consultations with Fermi, on September 21, 1929, he delivered a speech titled "The New Goals of Experimental Physics," which turned out to be prophetic in several aspects. I quote here excerpts from it:

A field of study whose theory is behind the times is that devoted to the mechanism of molecular or atomic arrangement of liquids and solids. It has been established that the forces of molecular cohesion are of electrical origin. Examination using x-rays has taught us almost to see the arrangement of the atoms or groups of atoms in crystal order. But the numerical prediction of physical constants of the atomic-electron ensembles is barely beginning. Hence there is a lot for theoretical physicists to do in this field. Also, on the experimental side, the field is anything but exhausted. Let me cite an old unsolved problem. If, in the structure of those ensembles, one could bring about some of the modifications that occur spontaneously in nature, we could, for example, transform coal or graphite into diamonds, which would not be devoid of interest, even for science. Physics dealing with the solid or liquid states of matter, or dealing with the effects of high pressure and very high or low temperature, must therefore be considered a promising field for today's or tomorrow's theoretical and experimental physicist. In addition, the applications are of the utmost practical importance.

We come now to the highest category in physics research: the discovery of new phenomena, such as the electric current and its various effects, x-rays, and radioactivity, which were once new. After much consideration, I do not hesitate to assert an opinion that may seem over-daring. I believe that modern physics already possesses the basic knowledge of the possible phenomena that may develop or be produced experimentally on earth. Therefore, except for the field of artificial modifications of the atomic nucleus (to which I will shortly return), our descendants will not be able to participate in the revelation of great new discoveries in physics. They will not share the experiences of those who saw the birth of electrical science, or the development of optics, or the discovery of new radiations.

. . .

Therefore, the only possibility of great new discoveries in physics lies in the chance that one might be able to attack the internal nucleus of the atom. This will be the worthy task of the physics of the future.

... So Rutherford's experiments have given us the only possibility of artificial transmutation of chemical elements. However, the effect has occurred so infrequently that we have only been able to find individual cases of disintegration, atom by atom. It would take, therefore, thousands of centuries to gather an amount of hydrogen detectable by chemical means. Evidently Rutherford's method of attack, although the most energetic available to us today, is still inadequate for our needs. Will it be possible to attack the atom in some other way?

... Only technical and financial difficulties that are, in principle, surmountable, are in the way of this great project. The purpose is not only the transmutation of an appreciable quantity of chemical elements, but also the observation of phenomena involving colossal amounts of energy, which occur in some cases of fragmentation or reconstitution of the atomic nucleus.

The nucleus of various elements is composed as we have said before, of protons, or hydrogen nuclei, and electrons. However, in the combination of several protons, for example four to make the helium nucleus, the mass of the compound is slightly less than the sum of the masses of the four protons. This decrease in weight is called the mass defect of the nucleus. As required by the theory of relativity, this mass decrease must be accompanied by the release of an enormous amount of energy. Thus, in the formation of the helium nucleus from four protons, for every gram of helium formed, about a billion and a half large calories should be produced, that is, the equivalent of 2 million kilowatt hours. Naturally, the inverse phenomenon, that is, the breaking of one gram of helium into hydrogen nuclei, would require the use of as much energy. In these nuclear phenomena, the transcendental importance of which does not need to be emphasized, one would transform matter into energy and vice versa to the tune of 25 million kilowatt hours for every gram of transformed matter.

One can therefore conclude that while great progress in experimental physics in its ordinary domain is unlikely, many possibilities are open in attacking the atomic nucleus. This is the most attractive field for future physicists. To participate in the general movement, either in the present trend, or in the future directions I have indicated, it is indispensable for experimentalists to have a sure grasp of the results of theoretical physics, and also to have increasingly large experimental means. Attempting to do experimental physics without a working knowledge of the results of theoretical physics and without ample laboratory means is like trying to win a contemporary battle without airplanes and guns.

.   .   .

... Thus, even if physics were to move toward a saturation level, the study of its application to other disciplines such as biology, if conducted by true experts, masters of all the resources of modern physics, could bring results of the greatest scientific and practical value. Rather than having a superposition of techniques, it would be even better if one could attain a fusion of the biological attitude with the attitude of new physics in the same brain.

At the age of 50, Hermann von Helmholtz, the great naturalist of the last century, gave up his chair of physiology to go and teach physics in Berlin. The times permitted, or even encouraged, this daring break. It was crowned by the greatest success for him and for science. Today, the situation is reversed. Experimental physics is rapidly reaching a state of maturity and completeness that will be difficult to increase. On the other hand, it has all its tools ready to help other less advanced sciences.

Corbino expressed his thoughts concerning new directions in physics explicitly, but he certainly was not the only one to feel the beginning of a new era. At that time the new generation of physicists in many universities was turning to nuclear problems. The quasi-monopolistic position of Paris and Cambridge was shaken, and new names replaced those of Marie Curie and Rutherford. Walther Bothe, James Chadwick, and Fréderic Joliot approached the center of the stage. Furthermore, the United States, which had contributed relatively little to the development of quantum mechanics, materially increased its importance. New experimental centers formed in Europe too; for instance, in Rome, which already had a reputation as a theoretical center thanks to Fermi.

All these new departures came to an extraordinary fruition in 1932, a year that, by coincidence, saw a cluster of epochal discoveries, reminiscent in impact to those around 1895, which were described at the beginning of this book. The main discoveries, in chronological order, were the neutron; the hydrogen isotope of mass 2, deuterium; the positive electron or positron; the coming of age of accelerators, and soon thereafter beta ray theory and the discovery of artificial radioactivity. I believe that I heard of two of these discoveries at the same tea preceding a physics seminar at the University of Hamburg.

Figure 9.1 Walther Bothe (1891–1957) (left) and C. D. Ellis in Rome in 1931. In the background, from left to right, E. Amaldi, G. Placzek, and G. C. Wick. (Photo E. Segrè.)

The Discovery of the Neutron

The first event was the discovery of the neutron. It had a complex and dramatic history. Unlike many other discoveries, such as the x-ray, which happened in one evening, the discovery of the neutron took two years. It even has an important prehistory. Rutherford had repeatedly thought of the possible existence of a neutral particle of protonic mass. He conceived it as a hydrogen atom in which the electron had fallen into the nucleus, neutralizing its charge. Rutherford's speculations on the behavior of such a hypothetical particle had been expressed in his Bakerian Lecture of 1920, which, as we have seen, kept alive the possible existence of such a particle in the minds of his pupils.

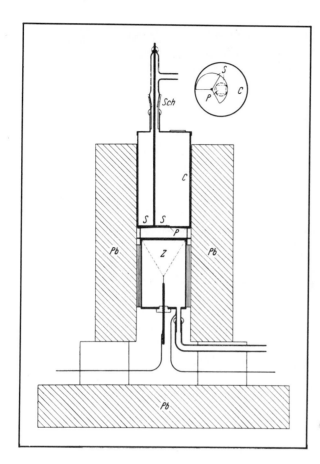

Figure 9.2 Schematic drawing of the apparatus used by Bothe and Becker with which they revealed penetrating γ radiation emitted from beryllium bombarded by polonium particles. Z denotes the Geiger counter [From *Zeitschrift fur Physik* 66, 289 (1930).]

Walther Bothe (Figure 9.1) and his student H. Becker took the first step toward the actual discovery of the neutron in 1928 when they bombarded beryllium with polonium alpha particles (Figure 9.2). Their aim was to confirm the disintegrations observed by Rutherford and to find out whether they were accompanied by the emission of high-energy gamma rays. Using electrical counting methods, they found a penetrating radiation, which they interpreted as gamma rays. They tried to measure their absorption coefficient in order to estimate their energy. They extended their observations to lithium and boron and concluded that the observed gamma rays had more energy than the incident alpha particles. This energy had to come from the nuclear disintegration. The investigation lasted a couple of years.

Walther Bothe (1891–1957), born in Oranienburg near Berlin, was one of the few pupils of Max Planck. He initiated his career at the Reichsanstalt in Berlin under Geiger, but during World War I he was taken

Figure 9.3 Frédéric Joliot (1900–1958) and Irène Curie (1897–1956). The young woman in Figure 2.8 is now an independent scientist and with her husband is preparing to carry out some great experiments. [F. and I. Joliot-Curie, *Oeuvres Scientifiques Complètes* (Paris: Presses Universitaires de France, 1965).]

prisoner by the Russians and sent to Siberia. There he studied mathematics and theoretical subjects, and married a Russian lady. At the end of the war he resumed his position at the Reichsanstalt and developed electrical counting methods. In particular he was the first to replace Rutherford's and Geiger's eyes, in their laborious scintillation counting, with an electric circuit, thus greatly enhancing the power of the method. The coincidence method was later applied by him and by his collaborators, Kohlhoerster, Rossi, and others to many problems in nuclear physics, cosmic rays, and the study of the Compton effect. Bothe was a first-class physicist—a real physicist's physicist, better known among physicists than by the public at large. His

Figure 9.4 Frédéric Joliot in a drawing by Pablo Picasso. (Courtesy of J. Hurwic.)

abilities extended to the arts, and he played the piano and painted at a professional level. He was somewhat difficult as a person, but his sterling intellectual and scientific integrity commanded great respect in the physics community.

Around 1931 two new scientists of great importance enter the scene: Irène Curie and her husband, Frédéric Joliot (Figure 9.3). Irène (1897–1956) was the daughter of Marie and Pierre Curie and resembled her mother in both character and physique; she had been reared with great care by her mother (her father died when she was nine years old). Marie and a group of scientists and intellectuals set up an elementary school for their children in which they taught—she of course, being the science teacher. During the First World War when Marie organized the radiological service for the French army, she took Irène as an assistant. With her chromosomes and such an education, it is little wonder that Irène turned to science, naturally in her mother's laboratory.

Frédéric Joliot (1900–1958) had been recommended to Marie Curie by her old friend Langevin, primarily for his exceptional technical ability. One of his first tasks was to prepare an extremely strong polonium source and subsequently to build a cloud chamber. Joliot (Figure 9.4) performed both of these tasks brilliantly, and furthermore married Irène, the boss's

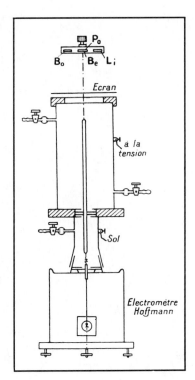

Figure 9.5 A schematic drawing of the apparatus used by Joliot and Curie for revealing the recoil protons caused by the neutral radiation emitted from Po–Be. On top, the source; in the center, the ionization chamber connected to an electrometer (below). [From *Comptes-rendus Academie des Sciences, Paris 194,* 273 (1932).]

daughter, in 1927. The new generation was thus well prepared to follow in the footsteps of the previous one. Joliot was cheerful, lively, cordial, and full of imagination; he reminded me of Maurice Chevalier, although my acquaintance with Joliot was slight and with Chevalier limited to films.

Joliot and his wife made all their great discoveries before World War II. After the fall of France Joliot entered the Resistance, and from 1941 to 1945 he was clandestine president of the National Liberation Front. After the war he became High Commissioner for Atomic Energy, but his politics, tending toward the extreme left, made him unfit for the job in the eyes of the French government and he was dismissed. He remained one of the major exponents of all international communist movements or initiatives, as well as director of the new scientific laboratory at Orsay. Irène became for a short period Minister for Scientific Research in the French Cabinet. Both died young, and it is probable that they were victims of their somewhat casual treatment of radiation and radioactive substances.

The Joliots decided to use their exceptionally strong polonium sample to study Bothe's penetrating radiation. On January 18, 1932, they reported a surprising observation of great import. The radiation was able to eject protons from a paraffin layer. They discovered this with an ionization chamber connected to an electrometer, but the result was so strange that

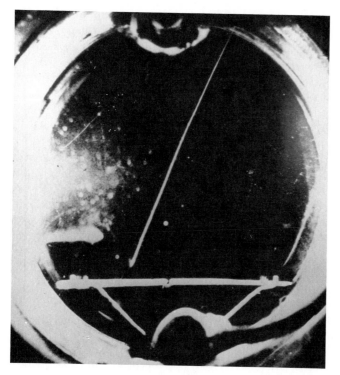

Figure 9.6 The track, in a cloud chamber, of a recoil proton set in motion by a neutron collision. The neutron does not leave a track because it is neutral. [From *Comptes-rendus Academie des Sciences Paris 194,* 847 (1932), Gauthier-Villars.]

they tried to confirm it immediately with a cloud chamber, and on February 22 they published the result of this second observation, confirming the ejection of protons (Figures 9.5, 9.6). Why was it so strange that Bothe's penetrating gamma rays should eject protons? The projection of a free particle by an impinging proton is a form of Compton effect well known for electrons. In the ordinary Compton effect, however, the recoiling electrons are light ($mc^2 = 0.51$ MeV) and recoil easily, but protons are 1,836 times heavier and do not recoil so easily. If one billiard ball hits another one, they easily recoil; but if an automobile is hit with a billiard ball, it will not be set in appreciable motion.

When Curie and Joliot tried to interpret their observations as a Compton effect, they made a most unlikely proposal in view of both: the energy that the incident "gamma rays" should have had and the collision cross-section that had to be attributed to them. This cross-section was about 3 million times larger than expected with a simple extension of the calculation valid for the electron. James Chadwick (1891–1974) reported to Rutherford the Curie-Joliot publication of January 18, and when his Lordship heard of the proposed explanation, it seems that he said with unusual vehemence, "I do not believe it." On reading the same paper, Ettore Majorana a young physicist in Rome, with his special sarcastic spirit, said, "What fools. They have discovered the neutral proton and they do not recognize it." Chadwick at the Cavendish did more. He repeated the experiments, using polonium plus beryllium as a source, but he collided the emerging radiation not only with hydrogen, but also with helium and nitrogen (Figures 9.7, 9.8). By comparing the recoils, he could then prove that the radiation contained a neutral component of mass approximately equal to that of the proton. He called it *neutron* and published the result by sending a letter to *Nature* on February 17, 1932. Curie and Joliot had thus missed a great discovery.

One of the reasons for Chadwick's speed and success was that he was mentally prepared for the concept of the neutron. He had previously made several attempts to produce the neutron in strong discharges and by other methods. In a paper on the discovery of the neutron he says that "some of these experiments were quite wildly absurd." To his great credit, when the neutron was not present he did not detect it, and when it ultimately was there, he perceived it immediately, clearly, and convincingly. These are the marks of a great experimental physicist.

It is rumored that Rutherford insisted that the Nobel Prize for the discovery of the neutron should go to Chadwick, who fully deserved it. To somebody who remarked to Rutherford that the Joliots had also made an essential contribution, Rutherford is said to have answered, "For the neutron to Chadwick alone; the Joliots are so clever that they soon will deserve it for something else."

The discovery of the neutron had vast and profound consequences for nuclear physics. Everyone up to 1930 had accepted the hypothesis that the nucleus was composed of electrons and protons (as Corbino said in the speech at the beginning of this chapter). The hypothesis seemed plausible because one saw electrons emitted by nuclei in beta decay, and the proton was the lightest known nucleus. Furthermore, the nuclear masses were approximately integral multiples of the proton mass. According to this hypothesis, the nucleus of nitrogen of mass 14 should contain fourteen protons and seven electrons, giving the mass of fourteen protons and neutralizing the electric charge of seven of them with seven electrons. The mass of the electrons is negligible compared with that of the protons, and one must also take into account the binding energy according to Einstein's equation $E = mc^2$. However, there were serious difficulties confronting this hypothesis. First, the uncertainty principle required a potential barrier of great magnitude to confine a particle as light as an electron in a volume of the size of the nucleus, and one did not know what forces could produce such a barrier. Even worse, some experiments in an entirely different field of physics, molecular spectroscopy, had shown beyond any doubt that the nitrogen nucleus must contain an *even* number of fermions. Protons and electrons are fermions, and their number in the nitrogen nucleus, according to the electron-proton constitution, should be twenty-one, an *odd* number. There was thus a grave clear-cut contradiction. This difficulty, and others concern-

Figure 9.7 James Chadwick (1891–1974). A student and colleague of Rutherford, Chadwick was one of the greatest nuclear physicists. His discovery of the neutron was basic to all the subsequent developments in nuclear physics. (Niels Bohr Library.)

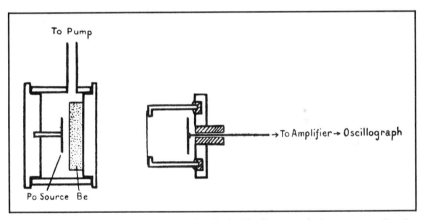

Figure 9.8 The apparatus with which Chadwick discovered the neutron. On the left, the Po–Be source; on the right, the ionization chamber connected to an amplifier. [From *Proceedings of the Royal Society, London* 136, 692 (1932).]

ing the spin, were all removed at once by the assumption that the nucleus was composed of neutrons and protons, both fermions: According to this theory, the nitrogen nucleus of mass 14 contains seven protons and seven neutrons, an even number of fermions. This hypothesis was put forward independently by D. Ivanenko in the Soviet Union and by Heisenberg in Germany. Majorana had reached the same conclusion but did not publish his thoughts.

The nuclear model based on a neutron-proton composition is now universally accepted and seems here to stay. To reach a complete understanding of this model it is then important to study the proton-proton, proton-neutron, and neutron-neutron interaction, with the ultimate hope of deriving all nuclear properties from such interactions. This ambitious program has not yet completely succeeded, in spite of significant partial successes. Experiments begun in the 1930s and extended and deepened over many years have demonstrated that specific nuclear forces between neutron and neutron, proton and proton, and proton and neutron are the same. Heisenberg, E. U. Condon, Eugene Wigner, and others gave this fact a theoretical formulation by considering the neutron and the proton as the same particle (nucleon) in two different quantum states differing by the value of an internal coordinate. This is perfectly analogous to an electron having spin parallel or antiparallel to a given direction fixed by a magnetic field. The internal coordinate of the nucleon, because of its analogy with the spin, has been called isotopic spin and is often denoted by the symbol T. The isotopic spin must not be thought of in ordinary space, but in a special abstract space. Mathematically, it has quantum mechanical properties, such as commutation rules, identical to angular momentum. In its own space it is a vector with three components, the third component being connected to the electric charge by the equation

$$Q = T_3 + \frac{N}{2} + \frac{S}{2}$$

where Q is the electric charge (1 for the proton), N is the number of nucleons contained in the system, and S is the strangeness, of which we shall speak later. For nuclei $S = 0$. Thus the discovery of the neutron removed many nasty difficulties and opened the way to a new understanding of the nucleus. It did not solve another fundamental problem besetting nuclear physics, that of beta decay, of which we shall also speak later.

The Discovery of Deuterium

The day after Chadwick sent his announcement on the neutron to *Nature,* the American *Physical Review* received another most important paper, reporting the discovery of a hydrogen isotope of mass 2. The authors were Harold C. Urey, then a professor of chemistry at Columbia University in New York; F. G. Brickewedde, head of the cryogenic section of the National Bureau of Standards in Washington, D.C.; and G. M. Murphy, of Columbia University.

Harold Urey (b. 1893) is the son of an Indiana clergyman. His father died when Harold was six years old, and his mother remarried, again to a clergyman. The family was very poor, and Harold had to support himself at an early age. He attended the State University of Montana, where he obtained a bachelor's degree in chemistry in 1917. He then took a position in a chemical plant engaged in war production, but as soon as possible he returned to the University of Montana as an instructor in chemistry. Only in 1921 did he arrive at a genuine research institution when he went to work for a Ph.D. at Berkeley under the brilliant Gilbert Newton Lewis (1875–1946). Lewis was primarily a physical chemist and a thermodynamicist, but his inquisitive and imaginative mind spanned a great variety of subjects in a stimulating, if not always critical, fashion. He dominated the Berkeley chemistry department, which was to a large extent his own creation. Lewis had a number of distinguished pupils in very diverse fields connected with chemistry; Urey was one of them.

After completion of his doctorate Urey went to Copenhagen with the help of a fellowship. There he learned atomic physics as known in 1924, and he wrote a book with A. E. Ruark that was influential in propagating the new ideas among American chemists. In 1929 he became professor of chemistry at Columbia University, and there he discovered deuterium. During the Second World War he was in charge of isotope separation by the diffusion process for the Manhattan Project. After the war he went to the University of Chicago and ultimately to the University of California at San Diego. His postwar activity has been devoted mainly to planetary and lunar problems. Urey has an enthusiastic personality, always ready to support good causes, a tradition common in the United States.

The isotopic composition of the natural elements, and precision measurements of atomic masses of single isotopes, from the time of Aston had shown that not even single isotopes have integral masses. We assume here, as usual, an atomic weight unit such that the weight of C^{12} is exactly 12. The fractional values of the single isotope mass is due to the nuclear binding energy. That chemical atomic weights have fractional values, on the other hand, is due mostly to the fact that chemical "elements" are isotopic mixtures. For instance, chlorine has a chemical atomic weight of 35.46 and is a mixture of two isotopes of atomic weight 34.97 and 36.97. Already in 1919 Otto Stern had considered the possibility that hydrogen, with an atomic weight of 1.0079, might be a mixture of two isotopes. He investigated this possibility with his colleague M. Volmer; the result was negative. They had assumed that the whole difference between 1 and 1.0079 might be due to the admixture of a hypothetical new isotope of mass 2, which should have been present in an amount of the order of 1 percent. The idea is fundamentally correct but quantitatively wrong because deuterium atoms are present only in the amount of about 0.015 percent.

Later, very careful investigations on band spectra of some elements led to the discovery of rare isotopes in oxygen and nitrogen, and the case of hydrogen was then reconsidered by Urey and his colleagues, who enriched the rare isotope by a fractional distillation of liquid hydrogen and ultimately

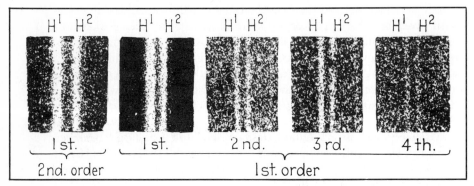

Figure 9.9 Lyman's spectral series of hydrogen isotopes. It shows the shifting of the deuterium lines in relation to the ordinary hydrogen ones due to the change in nuclear mass. The formula in Figure 7.3, which gives the frequency of the lines, contains the "reduced mass," which is different for deuterium and for hydrogen.

succeeded in demonstrating spectroscopically the presence of deuterium (Figure 9.9). Deuterium is an isotope of unusual importance in nuclear physics. By coincidence, one of the fundamental papers on the subject by M. S. Livingston and E. O. Lawrence appears back to back with the announce-and it even has practical applications in some nuclear reactors.

The wonder year of 1932 also saw the beginning of good performance by particle accelerators, which were destined to transform nuclear physics. By coincidence, one of the fundamental papers on the subject by M. S. Livingston and E. O. Lawrence appears back to back with the announcement of the discovery of deuterium and was received by the *Physical Review* on February 20, 1932, two days after Urey's paper and one week after Chadwick's announcement of the discovery of the neutron.

The Positron

The excitement generated by these discoveries had not yet abated when a new marvel appeared, this time from the field of cosmic rays. Cosmic rays were known from the beginning of the century. Their extraterrestrial origin was established around 1912, chiefly through the balloon flights of V. F. Hess (1883–1964) who, by measuring the spontaneous discharge of electroscopes at various altitudes, was able to distinguish the ionization produced by radioactive substances contained in the earth from that produced by penetrating radiation of extraterrestrial origin. Other admirable studies in physics and geophysics contributed to the conclusion that an isotropic radiation pervaded the whole universe. It was surmised that this radiation consisted of electrons and high-energy gamma rays. Any analysis and interpretation were faced, however, with serious difficulties because the radiation observed on the surface of the earth is not the primary radiation coming from space, but rather the products formed by the primary radiation in crossing the terrestrial atmosphere. High-altitude observations on mountains or in balloons could help to disentangle the primary from the secondary radiation, but the problem remained very complicated.

A very active school of cosmic ray investigators flourished around Robert Andrews Millikan in Pasadena, California. He had developed admirable techniques and had organized scientific expeditions and research groups that had greatly contributed to the study of cosmic rays. However, not all of Millikan's ideas were correct, and sometimes he was opinionated and unwilling to accept evidence contrary to his preconceived ideas.

R. A. Millikan (1868–1953), like so many American scientists, was the son of a poor clergyman with a large family. His youth was spent in midwestern small towns. He attended Oberlin College, where, as a sophomore, he was asked to teach physics, which he had never studied before. He continued teaching until, with a fellowship, he was able to go to Columbia University, where he obtained a doctorate in 1895. After a period in Germany he became assistant to A. A. Michelson in Chicago. Only in 1909 was he able to start a major research program on the determination of the charge of the electron, and later on Planck's constant, both first-class experimental accomplishments. From Chicago he moved in 1921 to the California Institute of Technology in Pasadena, where he established a great physics laboratory—for some time the best in the United States. Millikan's characteristics were unusual among physicists. He was an excellent public relations man, able to raise large sums of money even in difficult times. He did not hesitate to advertise himself; he liked to speak about science and religion, and referred to cosmic rays as "the birth cries of atoms" or "music of the spheres." Thus when a religious group put up a big sign at CalTech with the words "Jesus saves," some students added the phrase "And Millikan takes the credit."

Millikan's self-publicizing, however, had a very solid basis in scientific and administrative accomplishment. His own experimental research in physics was extended to other fields by the many outstanding CalTech groups. Cosmic rays, spectroscopy, x-rays, and nuclear physics thrived. Under his administration laboratories for chemistry, biology, engineering, and other disciplines were created and staffed with promising men of science, many of whom later became world famous. The Mount Wilson and Palomar observatories were leaders in astronomy.

Millikan's origins and background are reflected in his personality. He grew up in a rural atmosphere in the Midwest during a period that has been well described by Mark Twain. Millikan's versatility is evident from the varied work he undertook, from court recorder to physical education coach to an outstanding experimental scientist. His simple and occasionally naive

Figure 9.10 Carl D. Anderson (right), discoverer of the positron, and his student Donald Glaser, inventor of the bubble chamber, around 1950. (Photo by E. Segrè.)

views on religion and philosophy and his scientific style, in which technique and instrumental ability are more important than theoretical ideas, are all typical of the America of his time. Despite his weaknesses and simplicity, he had his own high ethical ideals and a certain nobility.

One of the most important of Millikan's protégés was Carl D. Anderson (Figure 9.10), born in 1905 in New York of Swedish parents. Anderson has spent his long and distinguished career entirely at the California Institute of Technology. In 1930 he started building cloud chambers subject to a magnetic field and containing metal plates through which the observed particles had to pass. He could thus observe their magnetic deflection and their behavior in crossing the plates.

Using this technique and its improvements or variations, Anderson and his pupils (S. Neddermeyer and others) made many important discoveries over a long period. One of the first occurred on August 2, 1932. On that day he obtained the picture reproduced in Figure 9.11, which shows an electron crossing a lead plate and stopping in the cloud chamber. The direction of motion is certain because in crossing the lead plate the electron loses momentum and, as a consequence, the trajectory's curvature increases. The electron hence must travel upward. However, given the direction of motion and the direction of the magnetic field, the curvature belongs to a positively charged particle, not to an ordinary electron, which is negatively charged. Is it a proton? Impossible, because the trajectory of a proton of sufficient momentum to cross the lead plate would not show any visible curvature in the magnetic field of the chamber; conversely, if its trajectory is curved, it

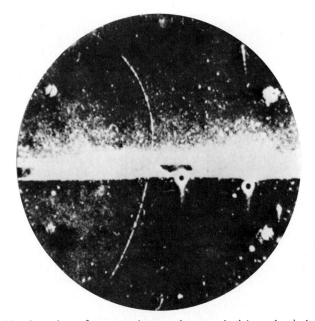

Figure 9.11 A positron from cosmic rays photographed in a cloud chamber subject to a magnetic field. The radius of curvature of the trajectory, the direction of the magnetic field, and the velocity of the particle form a left-handed coordinate system if the charge is positive. The positron comes from below and crosses the lead plate, where it is slowed. The decreased velocity is shown by the greater curvature of the trajectory in the upper part of the photograph. [Photo by C. D. Anderson, from *Physical Review 43,* 491 (1933).]

would have insufficient momentum to cross the lead plate. A detailed examination of the picture leads to the unavoidable conclusion that we are really dealing with a positive electron, or *positron,* as we now call it. In 1932 Anderson was unfamiliar with Dirac's theory of the electron and did not know that Dirac had predicted the positron. He published a short paper of such convincing experimental evidence that he immediately persuaded all physicists that he had observed a positive electron. Of course, this was a triumph for Dirac's theory, too.

Following Anderson's discovery, other physicists discovered positron tracks in pictures they had taken previously; at the time the tracks had escaped their attention, had been misinterpreted, or had not been sufficiently convincing to prove the existence of the positron.

Some time before Anderson's work Bruno Rossi had observed coincidences between three Geiger-Mueller counters placed in a triangular arrangement surrounded by lead. The coincidences were difficult to explain without assuming the simultaneous presence of at least two charged particles

Figure 9.12 A shower of electrons and positrons produced by cosmic rays. The shower is caused by the cosmic rays' creation of electron-positron pairs. The picture was taken by P. M. S. Blackett and G. Occhialini in a cloud chamber controlled by a Geiger counter. [From *Proceedings of the Royal Society 139*, 699 (1933).]

emerging from the lead. The results of the experiment appeared so incredible that the first scientific journal to which they were submitted for publication rejected the manuscript, which was then sent to another journal. In order to clear up the mystery, P. M. S. Blackett and G. Occhialini triggered the expansion of a cloud chamber using a triple coincidence between counters that surrounded it. They then saw a true shower of positive and negative particles in the cloud chamber (Figure 9.12). The work was performed at the Cavendish Laboratory. A story of the discovery, perhaps somewhat embellished, has it that when Occhialini saw the picture he had developed, he rushed to Rutherford's house. The maid opened the door and in his enthusiasm Occhialini kissed her. When Rutherford saw the picture, he wrote a check for fifty pounds to Occhialini, who was in financial straits at the time.

The shower phenomenon is explained by the materialization, as it is called, of gamma rays into electron-positron pairs, which produce more gamma rays, and so on. The materialization can occur only in the presence of nuclei, preferably heavy, which are needed to assure the conservation of energy and momentum in the process.

Irène Curie and Joliot had seen positrons in a cloud chamber even before Anderson when they were studying the radiation produced by polonium plus beryllium. However, they interpreted them as electrons moving toward the source rather than as positrons coming from the source. Of course, it was a problem to understand where the electrons moving toward the source came from ... Curie and Joliot must have regretted missing not only the neutron but the positron as well. In April 1933 they again started operating their cloud chamber, and on May 23 they demonstrated that hard gamma rays from a polonium plus beryllium source (which are emitted by it besides neutrons) produce electron-positron pairs by materialization. Two months later, in July, they also noted single positrons in addition to pairs. This was a remarkable fact and, even more remarkable, the energy of the single positrons seemed to have a continuum of values.

The New Nuclear Physics

The seventh Solvay Council from October 22 to 28, 1933, was devoted to the nucleus, which was of central interest in physics. As all the discoveries mentioned in this chapter were new, there was no dearth of subjects for discussion. Figure 9.13 shows the participants to the Council; the older generation is represented among others by M. Curie and Rutherford; the younger one by Joliot, Chadwick, Irène Curie, Bothe, Fermi, and E. O. Lawrence, who was the only American. A reading of the council proceedings shows the state of knowledge, the doubts, and the confusion then prevailing. Some problems were under control or had solutions in sight; others were in total disarray.

One of the most distressing puzzles was beta decay, which seemed to question nothing less than the conservation of energy. The electrons of beta decay have a continuous energy spectrum, yet the initial and final states are perfectly defined. Where does the energy go if the disintegration electrons do not have an energy equal to the energy difference between initial and final state? For years this problem had been under investigation. Physicists had checked, for instance, whether gamma radiation could compensate for the missing energy. Many experiments, even with calorimetric methods, failed to reveal the whereabouts of the missing energy. Furthermore, there were other difficulties confronting the conservation of angular momentum

Figure 9.13 The participants at the 1933 Solvay Council. In this photograph are gathered the greatest exponents, young and old, of nuclear physics of the times. Seated, from left to right: E. Schrödinger, I. Joliot-Curie, N. Bohr, A. Ioffe, M. Curie, P. Langevin, O. Richardson, E. Rutherford, T. DeDonder, M. de Broglie, L. de Broglie, L. Meitner, J. Chadwick. Standing: E. Henriot, F. Perrin, F. Joliot, W. Heisenberg, H. Kramers, E. Stahel, E. Fermi, E. Walton, P. Dirac, P. Debye, N. Mott, B. Cabrera, G. Gamov, W. Bothe, P. Blackett, M. Rosenblum, J. Herrera, E. Bauer, W. Pauli, M. Cosyns, J. Verschaffelt, E. Herzen, J. Cockcroft, C. Ellis, R. Peierls, A. Piccard, E. Lawrence, L. Rosenfeld. (Institut Solvay.)

and of other quantities usually conserved. In despair, Bohr was ready to abandon conservation of energy in nuclear phenomena. Pauli, who was more conservative, had proposed since 1930 that the electrons in beta decay were accompanied by light neutral particles that escaped observation but reestablished energy conservation and other conservation laws by taking away in an undetectable fashion the missing energy. In Rome Pauli's light neutral particle was usually called by the name *neutrino* (small neutral one, in Italian), as contrasted with Chadwick's *neutrone* (big neutral one, in Italian), and the former name was universally adopted when, at the end of 1933, Fermi formulated a true theory of beta decay based on the neutrino.

In speaking about beta spectra at the Solvay Council, Francis Perrin, the son of Jean Perrin, made a remark of great import. He commented that the positrons that appeared in the cloud chamber pictures of Curie and Joliot, with their continuous energy distribution, reminded him of beta decay spectra. He did not know how right he was.

Another surprising result discussed at the conference was that recently obtained by Otto Stern and his collaborators I. Estermann and O. Frisch at the University of Hamburg. They had repeated the Stern-Gerlach experiment on hydrogen molecules and had succeeded in measuring the nuclear magnetic moment of the proton, not that of the electron (Figure 9.14). The experiment was very difficult because the expected proton magnetic moment was three orders of magnitude smaller than that of the electron, as we can see by observing that the expression of the Bohr magneton contains the mass of the particle in the denominator. A simple-minded generalization would thus suggest that the ratio of the magnetic moment of the proton to that of the electron is the inverse of the mass ratio: $1/1,836$. This was the expectation of theoretical physicists, including the great Pauli, who was also at Hamburg then. Pauli told Stern that if he enjoyed doing difficult experiments, he could do them, but that it was a waste of time and effort because the result was already known. Following a seminar on the experiment then in progress, Stern persuaded all those present to write the result they expected over their signature on a piece of paper, and he stuck the sheet in his pocket. At the end it turned out, surprisingly, that the magnetic moment of the proton was about three times larger than the naively expected value. This is one of the first indications of the complicated structure of the proton.

Following the Solvay Council the participants returned home cogitating on what they had heard in Brussels. Within a few months new great discoveries followed.

Figure 9.14 Otto Stern's laboratory, Hamburg, 1931. O. R. Frisch is shown with the apparatus for measuring the magnetic moment of the proton. Frisch was a nephew of Lise Meitner and collaborated with her in the interpretation of the Hahn and Strassmann experiments on uranium fission. (Photo E. Segrè.)

Fermi kept meditating on beta decay. He liked the qualitative idea of the neutrino, but it had to be transformed into a quantitative theory in accordance with the rules of quantum mechanics. There was a difficulty: In the usual theories the number of particles is constant, while it is essential that electron and neutrino be created instantly at the time of the beta decay. It is true, however, that light shows a similar effect; one could say that light quanta are created at the instant an atom jumps from a higher state to a lower one.

Dirac, Pauli, and Heisenberg had attacked this problem in 1928, but their seminal papers on the subject, which initiated quantum field theory and quantum electrodynamics, were not easy to understand. They quantized the Maxwellian electromagnetic field in accordance with the rules of quantum theory as expressed through commutation relations; from this the existence of light quanta, as postulated by Einstein in 1905, followed mathematically. In 1929 Fermi, after glancing at a 1927 paper by Dirac on the quantum theory of radiation and seeing the results he had obtained, decided that he would deal with the subject in his own simpler way, avoiding the difficulties that were at that time obstacles to an understanding of Dirac. He then wrote a famous article on quantum electrodynamics that opened radiation theory to the rank and file of physicists. The article was rigorous and correct but avoided the use of creation and destruction operators that had been invented several years earlier by O. Klein, P. Jordan, E. Wigner, and others. These operators, as their name indicates, create or destroy particles; they differ for fermions or bosons and require a mathematical technique that Fermi had not mastered.

Returning from Brussels, Fermi decided that he now had to learn how to use these operators, and after some study he felt he knew the subject and was ready for an "exercise" that would check his proficiency. He tried to describe beta decay, modeling it as closely as possible to electrodynamics. In order to do this, he had to introduce a new fundamental interaction, that is, a new natural force such as gravitation and electricity. This so-called weak, or Fermi interaction requires a new universal constant, g, which could be determined from beta decay experiments. Fermi's theory explains the form of the beta spectrum, the mean life for beta decay, and many more characteristics of such decays. It is still valid today, with a fundamental addition made by the Chinese T. D. Lee and C. N. Yang, which will be mentioned in Chapter 12. Fermi's investigation of beta decay is probably his theoretical masterpiece and is of fundamental importance for particle physics. Fermi sent his manuscript to *Nature* at the end of 1933, and it was promptly rejected by the editor. It was soon thereafter published elsewhere.

At Rome we were studying beta ray theory when, at the beginning of 1934, we read a one-page communication by Curie and Joliot that stunned us.

What had happened in Paris? As mentioned, at the Solvay Council Curie and Joliot had reported a continuous positron spectrum emitted by some substances under alpha bombardment. They had pursued the investigation of this phenomenon, and their conclusions, dated January 19, 1934, were contained in a letter to *Nature* (Figure 9.15). As they wrote,

> Our latest experiments have shown a very striking fact. When an aluminum foil is irradiated on a polonium preparation the emission of positrons does not cease immediately when the active preparation is removed. The foil remains radioactive and the emission of radiation decays exponentially as for an ordinary radioelement. We observed the same phenomenon with boron and magnesium....

The one-page paper substantiated Rutherford's prediction that the disappointments suffered by Curie and Joliot would soon be offset by something big. Indeed, they had now achieved one of the most important discov-

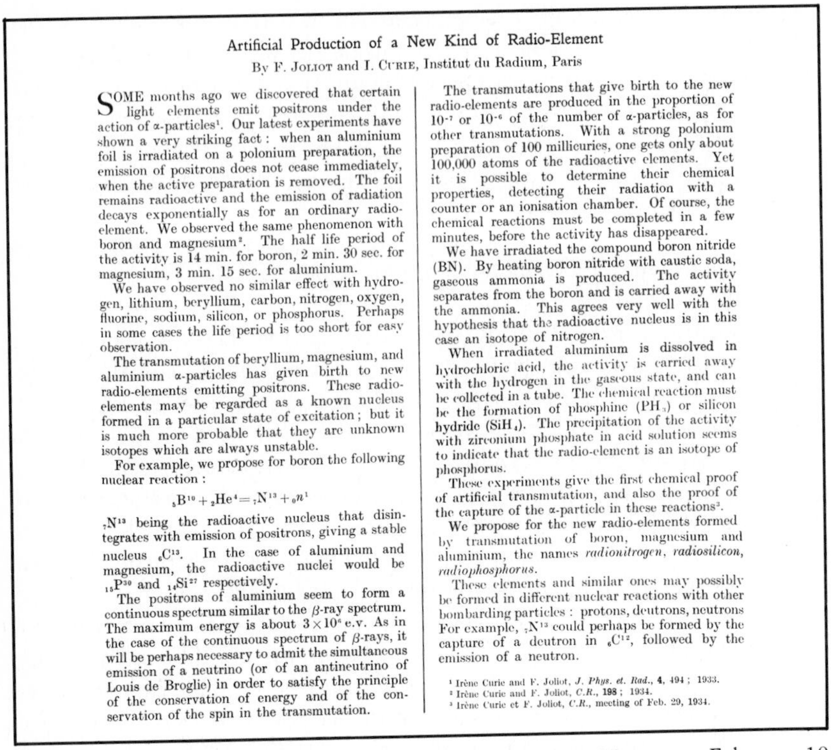

Artificial Production of a New Kind of Radio-Element

By F. Joliot and I. Curie, Institut du Radium, Paris

SOME months ago we discovered that certain light elements emit positrons under the action of α-particles[1]. Our latest experiments have shown a very striking fact: when an aluminium foil is irradiated on a polonium preparation, the emission of positrons does not cease immediately, when the active preparation is removed. The foil remains radioactive and the emission of radiation decays exponentially as for an ordinary radio-element. We observed the same phenomenon with boron and magnesium[2]. The half life period of the activity is 14 min. for boron, 2 min. 30 sec. for magnesium, 3 min. 15 sec. for aluminium.

We have observed no similar effect with hydrogen, lithium, beryllium, carbon, nitrogen, oxygen, fluorine, sodium, silicon, or phosphorus. Perhaps in some cases the life period is too short for easy observation.

The transmutation of beryllium, magnesium, and aluminium α-particles has given birth to new radio-elements emitting positrons. These radio-elements may be regarded as a known nucleus formed in a particular state of excitation; but it is much more probable that they are unknown isotopes which are always unstable.

For example, we propose for boron the following nuclear reaction:

$$_5B^{10} + {}_2He^4 = {}_7N^{13} + {}_0n^1$$

$_7N^{13}$ being the radioactive nucleus that disintegrates with emission of positrons, giving a stable nucleus $_6C^{13}$. In the case of aluminium and magnesium, the radioactive nuclei would be $_{15}P^{30}$ and $_{14}Si^{27}$ respectively.

The positrons of aluminium seem to form a continuous spectrum similar to the β-ray spectrum. The maximum energy is about 3×10^6 e.v. As in the case of the continuous spectrum of β-rays, it will be perhaps necessary to admit the simultaneous emission of a neutrino (or of an antineutrino of Louis de Broglie) in order to satisfy the principle of the conservation of energy and of the conservation of the spin in the transmutation.

The transmutations that give birth to the new radio-elements are produced in the proportion of 10^{-7} or 10^{-6} of the number of α-particles, as for other transmutations. With a strong polonium preparation of 100 millicuries, one gets only about 100,000 atoms of the radioactive elements. Yet it is possible to determine their chemical properties, detecting their radiation with a counter or an ionisation chamber. Of course, the chemical reactions must be completed in a few minutes, before the activity has disappeared.

We have irradiated the compound boron nitride (BN). By heating boron nitride with caustic soda, gaseous ammonia is produced. The activity separates from the boron and is carried away with the ammonia. This agrees very well with the hypothesis that the radioactive nucleus is in this case an isotope of nitrogen.

When irradiated aluminium is dissolved in hydrochloric acid, the activity is carried away with the hydrogen in the gaseous state, and can be collected in a tube. The chemical reaction must be the formation of phosphine (PH_3) or silicon hydride (SiH_4). The precipitation of the activity with zirconium phosphate in acid solution seems to indicate that the radio-element is an isotope of phosphorus.

These experiments give the first chemical proof of artificial transmutation, and also the proof of the capture of the α-particle in these reactions[3].

We propose for the new radio-elements formed by transmutation of boron, magnesium and aluminium, the names *radionitrogen*, *radiosilicon*, *radiophosphorus*.

These elements and similar ones may possibly be formed in different nuclear reactions with other bombarding particles: protons, deutrons, neutrons For example, $_7N^{13}$ could perhaps be formed by the capture of a deutron in $_6C^{12}$, followed by the emission of a neutron.

[1] Irène Curie and F. Joliot, *J. Phys. et. Rad.*, **4**, 494; 1933.
[2] Irène Curie and F. Joliot, *C.R.*, **198**; 1934.
[3] Irène Curie et F. Joliot, *C.R.*, meeting of Feb. 29, 1934.

Figure 9.15 The paper published by Curie and Joliot in *Nature* on February 10, 1934, announcing the discovery of artificial radioactivity obtained by bombarding certain nuclei (Al, B, and Mg) with alpha particles. The products decay, emitting positrons.

eries of the century: artificial radioactivity. They received the Nobel Prize, and I hope they felt compensated for having missed neutron and positron. At this point I might suggest to experimental scientists that it might be advisable to concentrate on short and juicy papers. The Nobel Committee seems to prefer one-page publications, in contrast to university promotion committees, which sometimes (although I believe this happens rarely) are supposed to judge by weight of paper!

The Joliots immediately verified the chemical nature of the new radioactive substances, separating them from the target by standard radiochemical means. They thus found that the nuclear reactions they had produced were

$$Al^{27}_{13} + He^4_2 = P^{30}_{15} + n^1_0$$

where the radioactive isotope of phosphorus P^{30}_{15} has a half-life of about 3 minutes and decays according to the reaction

$$P_{15}^{30} \rightarrow Si_{14}^{30} + e^+ + \nu$$

The upper and lower indexes following the chemical symbols are the mass number and the atomic number, Z. Nuclear reactions are often expressed in a more concise form introduced by Bothe. The formation of P_{15}^{30} would be expressed by the symbol Al_{13}^{27} (α,n) P_{15}^{30} where the first symbol denotes the target; the first in parenthesis, the projectile; the second in parenthesis, the particle that comes out; and the last symbol, the product of the reaction.

The consequences of the discovery of artificial radioactivity are immense. Very properly old Mme. Curie, by then mortally ill, could write to her daughter, "Nous voici revenus aux beaux temps du vieux laboratoire." [We have returned to the glorious days of the old laboratory.] She barely had the time to insert a short paragraph on artificial radioactivity in the new edition of her treatise on radioactivity, which was published after her death.

Chapter 10

Enrico Fermi
and Nuclear Energy

Enrico Fermi was born in Rome on September 29, 1901, the son of Alberto Fermi, an administrative employee of the Italian railroads, and his wife Ida de Gattis, a former school teacher. Fermi's grandfather had still tilled the soil near Piacenza in northern Italy, but through hard work and extreme thrift the family had improved its financial condition, and at the time of Enrico's birth they had a modest but secure livelihood.

Enrico grew up in Rome and attended high school there. He was a model student, first in everything. As a child, he discovered his own great interest in mathematics and physics. At the age of ten, after hearing in a conversation between adults that the equation $x^2 + y^2 = r^2$ represented a circle, he succeeded in discovering by himself the meaning of the statement.

Fermi had a slightly older brother, who was considered the more brilliant of the two. At age fifteen this brother unexpectedly died in a tragic surgical accident, and Enrico, who was very close to him, suffered a grave shock. After a very sad period he became friends with one of his brother's schoolmates, Enrico Persico. They remained intimate for life and were destined to be the first two professors of theoretical physics in Italy.

Fermi's father was friendly with another employee of the Italian railroads who had considerable technical knowledge, Adolfo Amidei. Amidei first met Enrico when he was about fourteen years old and soon recognized his uncommon abilities. He loaned the boy his own mathematics and engineering books and guided him in his readings. The young Enrico rapidly acquired a solid mathematical education by studying books on algebra, analysis, and geometry and by solving innumerable problems, many of which were far from simple, according to his mentor. When Enrico finished high school, Amidei suggested that he compete for a free university education at the Scuola Normale Superiore in Pisa. Fermi agreed and easily placed first in the admission competition. The archives of the Scuola Normale preserve his admission test.

The subject of the essay was "Characteristics of Sound." After a half-page introduction, the candidate proceeded to treat the example of a vibrating rod in detail. He wrote the partial differential equation for the rod, found its eigenvalues and eigenfunctions, developed its motions by Fourier analysis, and so on. Few Ph.D. candidates at the time, and even today, could have developed a treatment of such sophistication, from scratch, without any help from books. Furthermore, there were no errors at all. The examiner, Professor Pittarelli, was astounded; he decided to interview the candidate, although an oral examination was not prescribed. After talking to him, Pittarelli told Fermi that in his long teaching career he had never seen anybody like the young student, that Fermi was sure to win the competition, and that everything pointed to a brilliant career. Fermi told me this with much gratitude toward Professor Pittarelli.

At the Scuola Normale Fermi studied alone; his teachers were the books he found in the library. In letters to his friend Persico, Fermi reported his progress in detail. After one year, according to his statements to Persico (and he was not the bragging type), Fermi was considered the supreme authority on relativity and on quantum theory at the University of Pisa. Fermi acquired another friend in Pisa, Franco Rasetti, of whom I will speak later. Immediately after receiving his doctorate in 1922, Fermi returned to his family in Rome and there called on the director of the Physics Institute at the university (Figures 10.1, 10.2). Here is Fermi's own description of the meeting:

Figure 10.1 Orso Mario Corbino, physicist and politician (right). With Corbino are A. Sommerfeld (center) and R. A. Millikan (left). It was Corbino who obtained a chair in theoretical physics for Enrico Fermi at the University of Rome and fostered the development of a physics center in Rome. (California Institute of Technology.)

Figure 10.2 The Physics Institute at Rome, located on the Via Panisperna.

I became acquainted with Senator O. M. Corbino on my return to Rome, immediately after my graduation in 1922. I was then twenty years old; Corbino was forty-six. He was a Senator of the Kingdom, had been Minister of Public Instruction and was furthermore universally known as one of the foremost personalities in the world of scholarship. It was thus with understandable trepidation that I introduced myself to him, but my trepidation soon disappeared when he began discussing my studies with me in a cordial and interesting manner. In that period we had almost daily conversations and scientific discussions that helped not only to clarify many ideas that were unclear to me, but also to generate in me a deeply felt veneration of a student toward his master. This veneration steadily increased during the years I was privileged to work in his laboratory. I believe that these feelings are common to everybody who approached him. His cordiality, the clever and witty way in which he could express even unpleasant truths without hurting in the least, his complete sincerity and the real interest he had for scientific as well as human problems, elicited for him immediate liking and admiration. [Fermi, *Collected Papers,* vol. 1, p. 1017 (University of Chicago Press, 1962)]

Discoveries at Rome

Corbino had a dream: to resurrect physics in Italy. The science that he loved so well had been in the doldrums in Italy for almost a century after the period of great splendor it had enjoyed at the time of Volta and Avogadro.

Corbino, clever, big-hearted, and completely devoid of jealousy as he was, immediately saw in Fermi the means of fulfilling his dream. Hence he encouraged, protected, and helped him indefatigably. With a fellowship from the Italian Government, and later with a Rockefeller Fellowship, Fermi went to Germany and to Holland, where he came in contact with the international community of physicists. He thus emerged physically from the provincial Italian shell from which he had already escaped mentally during his student days at Pisa.

In Germany Fermi spent some time with Born at Göttingen, where he met Heisenberg and Pauli, but that stay, all told, was not very successful because Fermi remained very isolated. He had better luck at Leyden, where Ehrenfest recognized his ability and greatly encouraged him. On his return to Italy, Fermi obtained a temporary teaching job at Florence. After a minor failure in a competition for a chair in Sardinia, Corbino managed to secure a chair in theoretical physics for the University of Rome. It was the first in Italy for this subject, and Fermi was appointed to it. At the same time, Persico went to Florence to another newly created chair for theoretical physics. By then, in 1927, Fermi had acquired an international reputation, having discovered Fermi's statistics, which is applicable to particles obeying Pauli's exclusion principle.

In Chapter 5 we saw how Bose and Einstein had shown the changes needed in classical statistical mechanics to account for the sameness of the particles under consideration. Specifically, Bose applied these ideas to light quanta, and Einstein to molecules. However, their method of counting did not consider the restrictions imposed by the exclusion principle, since it was then unknown. Bose-Einstein statistics are valid for particles that are not subject to such a restriction; all these, called bosons, have integral spin. Particles with half-integral spin are subject to the exclusion principle, and for them one must use Fermi's statistics. Such particles are called fermions. Because electrons, protons, and neutrons, among others, are fermions, Fermi statistics has a very wide application; for instance, it is fundamental to the study of metals.

Once settled in Rome, Fermi, always with Corbino's help, secured an appointment there for Franco Rasetti (b. 1902) too, and both started assembling a promising group of students. In order of time: I, Ettore Majorana, Edoardo Amaldi, and others, all eager and able to learn, joined the small band. At the beginning the school in Rome worked chiefly in optical spectroscopy and atomic theory, but when Fermi and his friends perceived that new subjects were more promising and the future lay in nuclear physics, they changed direction. This change, of course, required time and effort, but in the end it paid off handsomely.

We have seen how, at the end of 1933, Fermi succeeded in solving a fundamental theoretical problem—beta decay. This work was most important, not only for its own results, but also because it served as an inspiration and model to H. Yukawa for his theory of strong interactions, which we will discuss in Chapter 12. Fermi's theory of beta decay has stood the test of time and, in fact, appears to be of increasing importance with passing years.

Rome acquired a reputation and became an important world center for experimental physics, as well as theoretical physics. Here is how it happened:

An improvement was needed for the full exploitation of the great discovery of artificial radioactivity: the use of neutrons as projectiles. In their pioneering experiments, Curie and Joliot obtained about one disintegration for every million alpha particles impinging on aluminum. The main reason for this low yield is that the aluminum nucleus repels the alpha particles by electrostatic action and prevents them from coming in contact with the target nucleus. It occurred to Fermi that in the case of neutrons there is no such electric repulsion, and the yield must approach unity. On the other hand, neutron sources of natural radioactive alpha emitters and beryllium emit very few neutrons per alpha particle because neutrons emerge only when a disintegration occurs in the source. Beryllium, however, has the atomic number 4, and hence it is easily penetrated. The use of neutrons gives, figuratively, the penetrability of beryllium to every element. In particular, for elements of Z larger than about 10, neutrons afford the only possibility of producing reactions with natural sources. These simple considerations induced Fermi to try to use neutrons as projectiles. This departure opened the way to completely unexpected developments.

At first, Fermi started irradiating all the elements he could set his hands on, in order of increasing atomic number. He had no success with hydrogen, lithium, beryllium, boron, carbon, nitrogen, or oxygen, but he was persistent, and finally fluorine showed radioactivity.

A period of intense, rapid work followed for the next three years. Fermi and his collaborators, Amaldi, O. D'Agostino, Rasetti, and I (later also B. Pontecorvo), at first discovered about forty new radioactive substances. This had great practical and theoretical importance. It provided a great increase in the material for studies on nuclear systematics, and in due course it gave radioactive tracers for practically all chemical elements, thus helping to revolutionize chemical and biological techniques.

In the spring of 1934 we irradiated the heaviest element then known—uranium. We found several radioactive periods and substances. For all other elements, we had observed one or more of the reactions (n,α), (n,p) and (n,γ). Furthermore, we showed by chemical means that none of the radioactivities produced in uranium could be ascribed to elements of atomic number greater than that of lead. We then thought that we had produced transuranic elements, for instance from U^{238} by an (n,γ) reaction U^{239}, which in turn by beta decay would produce 93^{239}. In this we were in error, at least in part; while it was true that transuranic elements were formed (including the reaction described above), what we had observed was something quite different. Our experiments were repeated and extended by Curie and Joliot in Paris, and by Hahn and Meitner in Berlin. They confirmed most of our results but, by extending the investigation, found more and more substances and were forced to consider ever more complicated possible paths of decay. It became increasingly difficult to make sensible hypotheses on the genetic relationships between the substances formed and to locate them at the end of the periodic system. We had the impression of being confronted with a mystery, but of course we had no inkling of what it might be.

Our results were communicated in a guarded form. Fermi consistently refused to name the new hypothetical elements because he felt uneasy about the interpretation of the experiments. Fermi was a member of the Italian Academy, a creation of fascism, and by 1934 he was prominent in Italy even beyond scientific circles, although he was personally averse to any form of publicity. I believe there were suggestions, if not outright pressures, to add glory to the fascist regime by giving to the hypothetical new elements some name reminiscent of the fascists, such as Littorio. (The lictors were Roman officers bearing the fasces as the insignia of their office.) Corbino, who had an extremely prompt repartee, pointed out that the new elements had very short lives, and this might make them inappropriate to celebrate fascism. Only in 1938, in his Nobel speech, did Fermi put forward tentative names for the new elements. The moment was unfortunate; at that very time Hahn and Strassmann were discovering nuclear fission, thus proving that those elements were composed (to put it diplomatically) of poor chemistry.

In the fall of 1934, however, we faced a major surprise. We found, both by chance and by good observation, that neutrons filtered through paraffin were much more effective in producing nuclear reactions than those emerging directly from a radon plus beryllium source. Once the facts were ascertained, Fermi put forward the unexpected explanation that neutrons were slowed down by elastic collision in passing through paraffin and that slow neutrons were much more effective than fast ones in producing certain nuclear reactions. Within a few hours we were able to check this hypothesis, and on the evening of October 22, 1934, the same day in which we had discovered the effect, we wrote a one-page note signed by Fermi, Amaldi, Pontecorvo, Rasetti, and Segrè that firmly established the facts and their interpretation (Figure 10.3). For all the authors this was one of the masterpieces of their career. The great practical possibilities of slow neutrons did not escape Corbino, who insisted that we should take a patent on them, although of course at the time nobody could suspect that slow neutrons would be the key to nuclear energy.

We have reached 1935, the time of the Ethiopian War, the Spanish Civil War, and other forerunners of World War II. It was clear to informed and thinking people that the European situation was approaching a catastrophe. At about this time, for various reasons, the Rome group dispersed: Rasetti, alarmed and disgusted by the situation, came to America; I obtained

Figure 10.3 The authors of the research on radioactivity induced by neutrons. From the left, O. D'Agostino, E. Segrè, E. Amaldi, F. Rasetti, E. Fermi. Among the signers of the note written on October 22, 1934, was also B. Pontecorvo, missing in this photograph. (Photo E. Segrè.)

a physics chair at the University of Palermo in Sicily, relatively distant from Rome; Pontecorvo joined Joliot in Paris; Fermi and Amaldi remained alone in Rome. In 1937, unfortunately, Corbino died unexpectedly. The deteriorating general situation further forced the demise of our group. The enslavement of Italy to Germany by the formation of the Rome-Berlin Axis (Berlin-Rome when seen from Berlin), the promulgation of anti-Semitic laws in Italy, and the pursuit of other foolhardy policies rendered the situation untenable.

The Discovery of Fission

In the fall of 1938 Bohr told Fermi that he would most probably win that year's Nobel Prize. It was an unprecedented but purposeful breach of confidence, dictated by the exceptional circumstances prevailing at the time. Fermi then decided to leave Italy, and after receiving the Nobel Prize in Stockholm, he proceeded directly to New York with his wife and children. He was expected and welcomed with open arms at Columbia University, where he had previously visited.

Fermi had barely arrived at New York when Bohr, on his way to Princeton for a series of lectures, communicated the electrifying discovery of fission that Hahn and Strassmann had achieved on almost the exact days Fermi was in Stockholm for the Nobel ceremonies. Not a word of it had reached Fermi. Hahn and Strassmann had found radioactive barium among the products of neutron bombardment of uranium. They were stunned by the observation, but it was experimentally airtight and had to be believed. Here are their comments in the paper in which they announced their findings:

> As a consequence of these investigations we must change the names of the substances mentioned in our previous disintegration schemes, and call what we previously called radium, actinium, and thorium, by the names barium, lanthanum, and cerium. As nuclear chemists who are close to the physicists, we are reluctant to take this step that contradicts all previous experiences of nuclear physics. [*Naturwissenschaften 27*, 11 (1939)]

Hahn was, of course, the same Hahn we met as one of the first followers of Rutherford in Canada. Fritz Strassmann (b.1902) was Hahn's pupil; in spite of his hostility to the Nazis he had managed to preserve his job at the Kaiser Wilhelm Institute. Hahn was 59 years old when he discovered fission. He thus capped his brilliant career with his greatest discovery. He regretted, however, that Lise Meitner (1878–1968), with whom he had collaborated for many years, including the studies on the products of neutron bombardment of uranium, could not be there (Figures 10.4, 10.5). In spite of his great efforts and those of other important scientists to protect the gentle Lise from the fury of Hitler's racial hatred, she had to flee for her life. She had managed to stay and work in Berlin until Hitler annexed her native Austria. After the Anschluss she found herself in mortal danger. She fled Germany through Holland with the help of Dutch friends, and from there she proceeded to Sweden, where she found asylum. Hahn wrote to her of his astounding experiments as soon as he was sure of the facts. Meitner showed the letter during the Christmas vacation to her nephew Otto Frisch who was visiting her. (We have met Frisch previously as a co-worker of O. Stern.)

He too now was a refugee in Copenhagen. Frisch and his aunt were puzzling on what could be the explanation of Hahn's result. The facts seemed indisputable, and an explanation had to be found. It finally dawned on them that perhaps the uranium nucleus breaks up into two big fragments—it fissions, as they said. This phenomenon had been suggested as early as 1935 by a woman chemist—Ida Noddack in a note in which she criticized some of our own Rome experiments; she objected to the fact that we had not proved that uranium did not break into two big fragments. This paper was known to us, to Hahn and Meitner, to Curie and Joliot, and presumably at the Cavendish, as well as in other places; it had not been duly appreciated, however. Moreover, Noddack never bothered to do the relatively simple experiments that would have substantiated her hypothesis.

As soon as fission was understood, in January 1939, Frisch looked for and saw the large pulses produced by the fission fragments in an ionization chamber connected to a linear amplifier. This experiment clinched his in-

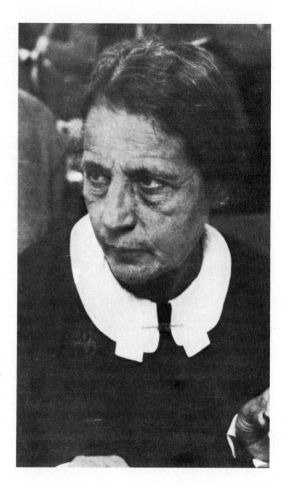

Figure 10.4 Lise Meitner (1878–1968) in 1937. She was Planck's assistant, then became a scientist at the Kaiser Wilhelm Institute in Berlin Dahlem and worked on many problems in nuclear physics, often with O. Hahn, before she was forced to flee the Nazis. She had a prominent role in the discovery of nuclear fission. (Photo E. Segrè.)

terpretation of Hahn and Strassmann's results. Many other physicists immediately verified the result. I know of at least four or five groups who, within days of the announcement of fission, had obtained experimental confirmation. One may ask why we did not discover fission in Rome. We had performed an experiment that came extremely close to a demonstration of fission. In order to see possible short-life alpha radioactivities produced by neutron bombardment in uranium, we assumed that such activities would give rise to energetic alpha particles. We covered our sample with an aluminum foil to stop these long-range alphas, and as a result we did not observe anything. If we had removed the aluminum foil, we would have seen the huge ionization pulses produced by the fission fragments. One wonders whether we would have interpreted them correctly. I have heard that in other laboratories the experiment had been performed and the con-

Figure 10.5 The fission of a uranium nucleus produced by neutrons. The two fragments leave a thin vertical sheet containing uranium, and their almost horizontal tracks can be recognized by little lateral branches caused by collisions. [From I. K. Bøggild, in *Physical Review* 76, 988 (1949).]

clusion drawn was that there was something wrong with the detector. Usually one recognizes only the expected, as can be seen in the examples of x-rays, the neutron, and the positron.

The Steps to the Atomic Bomb

The step from fission to a chain reaction is short in principle. The fission fragments necessarily have excess neutrons compared to stable nuclei of the same mass number. They can eliminate this excess either by the slow process of beta decay or by direct neutron emission if there is enough energy. In the second case the secondary neutrons may be used to produce new fissions, and if they are in sufficient number, they will produce more neutrons than in the first generation. In this way one achieves a divergent chain reaction. If the chain reaction occurs very rapidly in an uncontrolled way, a violent explosion follows and one has an atomic or, strictly speaking, a nuclear, bomb. On the other hand, if the reaction can be controlled and brought to a stationary state, one obtains a power source. Both paths lay open: to the atomic bomb and to nuclear power—a Janus-like opening—as often happens in the applications of science and technology. Many such ideas occurred to several physicists; some kept them to themselves, while others tried to take out patents. There was also an attempt by some physicists to keep the whole subject secret and to retard or impede its development, in fear of its possible consequences. This attempt faltered because of practical difficulties. Fermi decided to investigate fission experimentally and started work at Columbia University on a modest scale. He wanted to ascertain enough details of the fission process to decide whether a chain reaction was a daydream or a serious possibility. It would take too much space to give the details of this fascinating quest. It so happens that a chain reaction with natural materials, without any isotope separation, is a marginal process, at most barely possible. It is thus difficult to establish whether it can or cannot occur.

However, it was becoming quite clear that the possibility of nuclear explosives of unprecedented power was not a product of science fiction, but something to be taken extremely seriously, especially if one of the possible contenders for them was Adolf Hitler. Among the earliest physicists to study the problems of atomic energy, and to call attention to them, were the Hungarians L. Szilard, E. Wigner, E. Teller; the Austrian V. Weisskopf, and above all, Fermi. The singular phenomenon of a voluntary mobilization of many American scientists followed. Experiments were started, and secrecy was established on a voluntary basis. In August 1939 Szilard and other Hungarian activists who worried about the slow pace at which the work progressed planned a direct approach to President Roosevelt. Einstein was induced to sign a letter explaining the situation and its implications to the president. Fermi had undertaken similar steps a few months earlier by alerting representatives of the U.S. Navy, but the governmental machinery was slow, and the help it supplied at the beginning was puny. While this was occurring in the United States, similar efforts were started in the United Kingdom, in France, and in Germany.

In the meantime Bohr had presented theoretical grounds to suppose that U^{238} underwent fission only if bombarded with neutrons of an energy greater than about 1 MeV, whereas slow neutrons could produce fission only in the rare isotope U^{235}, which is present in natural uranium in the amount of about 1 part in 137. A. Nier and J. R. Dunning soon confirmed this hypothesis by experiment. It followed that in order to make a uranium bomb it would be necessary to separate the isotopes, a desperate enterprise apt to scare all but the most optimistic physicists.

Transuranic Elements

In 1940 E. McMillan and P. Abelson succeeded in identifying the first transuranic element, which they called *neptunium* (Np^{239}), and it was plausible that it should decay into a rather long-lived isotope of element 94. During a visit I made to New York around Christmas of 1940 Fermi and I talked at length of the properties to be expected of a nucleus of atomic number 94 and mass 239, later called plutonium 239—Pu^{239}. In particular, we were intrigued by its possible long life and by the likelihood that it would undergo fission under slow neutron bombardment. If such was the case, it could provide an alternative to U^{235} and avoid the necessity of separating isotopes. Of course, in order to prepare sizable amounts of the new substance, one had to build a nuclear reactor that could form it, and natural uranium might be suitable as fuel for such a reactor.

I had some experience with artificial chemical elements from the investigations of technetium and astatine (see p. 136) and I had worked on neptunium too, although I did not properly identify it.

In order to substantiate the feasibility of the plutonium approach, one had to produce a sufficient quantity of this new artificial element to ascertain its nuclear properties—in particular, how easily it would undergo fission under slow neutron bombardment or, in technical parlance, its fission cross-section for slow neutrons. If it turned out that one could use it instead of U^{235}, one had then to produce enough of it for a bomb. This was not a small enterprise, but the alternative of separating the isotopes was certainly tremendously difficult and expensive. In the first months of 1941 J. W. Kennedy, G. T. Seaborg, A. C. Wahl, and I prepared about 1 microgram of Pu^{239} with the Berkeley cyclotron and showed that it could be used as a nuclear fuel.

There were now two alternative ways of making an atomic bomb: the isotopic separation of uranium or the formation of a sufficient quantity of plutonium. Isotope separation could be achieved either by building a battery of large mass spectrographs, or by gaseous diffusion. The production of plutonium required first a nuclear reactor that would form it and then a chemical industry that could separate and purify it. There was no way of producing sufficient plutonium by nuclear bombardment using artificially accelerated charged particles.

Once either the explosive U^{235} or Pu^{239} became available, one still had to build the bomb proper.

Physics Mobilized

Before giving further information on the scientific aspects of the very complex enterprise that brought about nuclear energy and nuclear weapons, I will give a summary sketch of the administrative history. I have mentioned the early activities undertaken by the physicists, mainly on their own. It is noteworthy that the initiators were mostly Europeans. The scale of U.S. government support was small; $6,000 granted to Columbia University in February 1940, and $40,000 for 12 months from November 1940 to November 1941 were among the larger subsidiaries.

Early in the war the U.S. government and the most influential American physicists had been busy with radar and other urgent projects. They were skeptical about the practical possibilities of the release of nuclear energy, and immediate needs had first priority. Nuclear energy and explosives appeared too remote and too uncertain to warrant a concentration of scarce personnel and material resources in a time of crisis. In Great Britain, on the other hand, the scientific establishment pushed the nuclear effort even at the most critical moments of the war, and the example impressed many American scientific leaders who had continuous contacts with their British counterparts.

Serious government participation in the United States developed in 1940–1941. A number of committees were appointed, reconstituted, and dissolved, but by June 1940 the influential National Defense Research Committee (NDRC) emerged. It was headed by Vannevar Bush, a professor of electrical engineering at MIT, and an inventor. The President gave NDRC vast powers, among them jurisdiction over the uranium problem. J. B. Conant, an organic chemist and president of Harvard University, represented Bush in nuclear matters. Reviews by the National Academy of Sciences confirmed the importance and the possibilities of nuclear energy.

By the end of 1941 there had been a deep psychological change. In the words of the Smyth report: "Possibly Wigner, Szilard, and Fermi were no more thoroughly convinced that atomic bombs were possible than they had been in 1940, but many other people had become familiar with the idea and its possible consequences. Apparently, the British and the Germans, both grimly at war, thought the problem worth undertaking. ..." In short, it was decided to make an all-out effort. The controlling committee was streamlined and reorganized, always under the overall authority of Bush and Conant; other members were L. Briggs, formerly of the Bureau of Standards; H. A. Compton; Urey; Lawrence; and E. V. Murphree, from Standard Oil Development Company.

President Roosevelt was periodically informed of the progress and prospectives of the project. In 1942 the army was assigned an active part in procurement and engineering phases, and in September 1942 the Manhattan District was organized for the purpose. The Secretary of War placed General L. R. Groves in charge of the army's activities, and he became the director of the project. General Groves was a career officer who had distinguished himself in the construction of the Pentagon in Washington. He was an intelligent, honest, and courageous person, and an able administrator. Groves found himself in a completely new environment from those of his past, but he rapidly managed to recognize facts and people with good judgment (Figure 10.6).

It is difficult to give even a superficial idea of the complexity of the nuclear energy enterprise. It was necessary to create an entirely new technology with problems never encountered before on an industrial scale. The operation started in university laboratories—at Columbia, at Berkeley,

Figure 10.6 Sir James Chadwick and General L. R. Groves—different yet friendly (1944). (United Kingdom Atomic Energy Authority.)

at Chicago. Later special laboratories were created, on an intermediate scale, between the research and the production stage. These were at Argonne, Illinois; Oak Ridge, Tennessee; Richland, Washington; and other locations. As work progressed, major industrial companies were entrusted with development and operation, mostly under nonprofit contracts. This cluster of enterprises required coordination of geographically distant laboratories and plants, wrestling with entirely new technologies. Secrecy also had its requirements and had to be preserved as much as possible, creating additional difficulties for smooth operations.

Under normal circumstances all this would have required years, but in the hurry required by war demands, traditional stages were leapfrogged; for instance, traditional pilot plants were often omitted. Four years passed from the discovery of fission to the first critical reactor, which was started at Stagg Field in Chicago on December 2, 1942; similarly, four years lapsed between the discovery of plutonium and the first bomb. The individual laboratories had clear, well-defined assignments; once the goal was achieved, the laboratories were dissolved and the personnel reassigned to new tasks.

The Project as a whole was eminently successful. I will later discuss some of the reasons, in my opinion, for its success. Furthermore, from a financial point of view the expenditure was relatively modest: In round figures it cost 3 billion of 1940 dollars.

I will now return to the more technical aspects of the Project and to the personalities of some of the scientific leaders.

Fermi took charge primarily of the chain reaction necessary to produce plutonium. The problem was to assemble enough natural uranium and suitable neutron moderators to obtain a self-sustaining reaction, or *critical assembly* as it is called. This critical assembly would be a source of excess neutrons, which, when absorbed in U^{238}, would first form U^{239} and subsequently, by beta decay, Np^{239} and Pu^{239}. This last isotope has a half-life of 24,000 years and can be chemically separated from uranium. It is a nuclear explosive. The substances chosen for the reactors were ordinary uranium and graphite as moderator. It was difficult to obtain a chain reaction with them, but the scientists wanted to avoid any isotope separation or enrichment that, at the time, presented unsolved technical problems of staggering difficulty. The chief problem was to minimize neutron losses by parasitic absorption, geometrical leakage, and so on. In particular, maximum purity of the uranium and graphite was essential. On this occasion Fermi proved to be not only a great scientist but also an excellent engineer, and, as far as neutrons are concerned, as thrifty as his peasant forebears. The expertise on slow neutrons acquired in Rome was handy, and the theoretical study made in Italy, extended and modified to take into account the generation of neutrons by fissioning material, formed the theoretical underpinning for the building of the first reactor. Of course, Fermi was not alone in this work; many students and younger colleagues helped him valiantly, among them, H. L. Anderson, W. Zinn, L. Marshall, A. Wattenberg, B. T. Feld and others no less important. E. Wigner, A. Weinberg, and others were essential in forming the connection with industry.

After our discovery of Pu^{239} important chemical developments were needed to pass from the original microgram amounts to the kilograms of pure metal required for a bomb. The Metallurgical Laboratory (a code name) at the University of Chicago led in the first investigations of the process for extracting plutonium from irradiated uranium; industry, through the du Pont Company, managed pilot plants at Oak Ridge, Tennessee, and later production plants in the state of Washington. The metallurgy was accomplished at the Los Alamos Laboratory: G. T. Seaborg, J. W. Kennedy, and C. S. Smith were among the scientific leaders. Many engineers of the du Pont Company, Union Carbide, and other corporations were absolutely essential for success.

The separation of uranium isotopes by mass spectrograph was originally undertaken at Berkeley under the leadership of Lawrence, assisted by R. L. Thornton, W. Brobeck, and many others. Production moved to the Eastman Company in Oak Ridge.

The isotope separation by gaseous diffusion was led by J. R. Dunning, H. Urey, and others, and the plants at Oak Ridge were operated by the Kellex Corporation and the Carbide and Carbon Chemicals Corporation.

Once nuclear explosives were available, they had to be fabricated and assembled in a bomb. This planning and building of the bomb was a novel and far from simple enterprise. Furthermore, the studies on the bomb could not wait for the materials. The assembly method had to be ready by the time the materials were available. Preliminary theoretical studies on a small scale were carried out at Berkeley under the guidance of J. Robert Oppenheimer (1904–1967). A few months later, the need for a special laboratory devoted to the building of the bomb became clear, and Groves, by then in charge of the Manhattan Project, selected Oppenheimer as director.

Oppenheimer is one of the most discussed and controversial figures of the atomic age. Born in New York in a wealthy Jewish family of German origin, he was given the best in schools, teachers, and educational opportunities from his early youth. He was precocious, and the family, recognizing his talent, gave him every encouragement to develop it. He had rather eclectic tendencies and studied not only science but also many other subjects, such as philosophy, languages, and art, in a sophisticated but perhaps somewhat superficial way. His mind was extraordinarily quick and his memory remarkable, but he was also very conscious of his abilities and somewhat prone to arrogance, a weakness that made him many enemies.

He studied physics at Harvard University, and was among the first in America to grasp quantum mechanics. After his Harvard doctorate Oppenheimer worked as a postdoctoral fellow under Born at Göttingen and Bohr at Copenhagen. When he returned to the United States, he established thriving schools of theoretical physics at Berkeley and at the California Institute of Technology in Pasadena, spending part of his time in one place and part in the other. In his young years Oppenheimer was politically very naive and developed strong leftist tendencies that to me appeared romantic and completely uncritical.

Groves's choice of Oppenheimer may appear surprising, but it was certainly a felicitous one. Oppenheimer, who up to then had been a rather absent-minded and unworldly theoretical physicist, rose to the occasion and proved a superb laboratory director from both the technical and the administrative point of view. He also knew how to deal with people as diverse as Niels Bohr and General Groves.

The new laboratory site required isolation and a specific geographical location. Oppenheimer knew the plateaus of New Mexico from his boyhood and helped to select the site in a spectacular location where there had been a private school for boys, called the Los Alamos School. The site was rapidly organized and staffed with an uncommon assembly of scientific talent (Figure 10.7). The first three atomic bombs were built there.

After the war, for various reasons, Oppenheimer became very important in political and scientific circles. He aroused deep and implacable hatreds as well as extreme admiration and loyalties. He found himself at the center of controversies on matters of policy, and his enemies decided that his ruin was necessary for the salvation of the country. An inquest on his

Figure 10.7 A Sunday walk at Los Alamos. Standing, from the left, E. Segrè, E. Fermi, H. A. Bethe, H. Staub, V. Weisskopf; seated, Erika Staub and Elfriede Segrè. (Photo E. Segrè.)

loyalty was started under the Eisenhower Administration and ended unfavorably for Oppenheimer, who was declared a "security risk" and was barred from government service. The action elicited strong feelings in the scientific community, which divided into two camps in this issue. Years later, the American government, under Presidents Kennedy and Johnson, made amends for the unjust condemnation of a devoted servant by giving him the Fermi Award for nuclear work. By then, however, Oppenheimer had retired from political activity and returned to physics as Director of the Institute for Advanced Study in Princeton. He died there in 1967. The Oppenheimer case has inspired dramas, biographies, and a vast literature. I have indicated in the bibliography some of the writings that appear to me realistic and not utterly fictional.

In the creation of the new technologies there were occasional crises that temporarily upset the enterprise and even cast doubts on the outcome. I

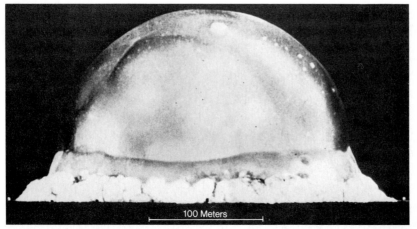

Figure 10.8 The first atomic explosion, the "Trinity Test," on July 16, 1945, at Jornada del Muerto, near Alamogordo, New Mexico. A fortieth of a second has passed since the start of the explosion. (Los Alamos Scientific Laboratory.)

was personally involved in one of these crises when my group, including O. Chamberlain and C. Wiegand—then students—was studying the spontaneous fission of Pu^{239}. We found that the Pu^{240} isotope, which was formed in reactors, together with Pu^{239} underwent spontaneous fission at such a high rate that it would have precluded the assembly of a bomb by the means then considered: The bomb would have predetonated and fizzled. It was a serious predicament; without a remedy half of the project would become useless for war purposes. The remedy was found, but it required true new inventions in which Neddermeyer, the former pupil of C. D. Anderson, had a central part.

Another crisis occurred at the start of the first production reactor. Some fission products turned out to absorb neutrons at an unexpected high rate, and thus they stopped the chain reaction. They "poisoned" the reactor, which could resume work only after the products had spontaneously decayed, to be stopped again after they had formed once more. This difficulty was overcome because the engineers had conservatively provided for the possibility of enlarging the reactor and increasing its reactivity in case of some unexpected event. The physicists had not liked the conservative attitude of the engineers that "unnecessarily" increased costs and wasted materials, but the engineers were amply vindicated. These crises are only two examples of many.

The original purpose of Los Alamos was fulfilled with the detonation of the first atomic bomb at dawn of July 16, 1945 (Figure 10.8). According to the U.S. Constitution, the president, as commander in chief of the armed forces, had the final say on the use of the bomb. President Truman consulted

Figure 10.9 The scenery around Los Alamos in 1943. Today Los Alamos is the seat of an American National Scientific Laboratory. (Photo E. Segrè.)

with several advisory groups, including one of scientists, and reached his conclusions with a full understanding of the alternatives and consequences. The wisdom of his decision has been debated ever since; however, it is hard to imagine that he could have chosen a different course of action at the time.

This is not the place for a discussion of the military and political events that followed. After the war, Los Alamos remained as one of the major national laboratories of the United States, with assigned goals for both war and peace, but the laboratory and the city of Los Alamos as they were during the war do not exist any more (Figure 10.9). They have been replaced by a more or less conventional city in which government architects and planners have ruthlessly destroyed the unique natural beauty, as is common in the building of many cities. Los Alamos lives in the memory of many of the early participants as a remembrance of a unique youthful and romantic period of their lives.

Consequences of the Bomb

The speed and success with which atomic energy and atomic weapons were developed has left many people with the belief that such colossal enterprises can be duplicated almost at will, given the money. This opinion is much too optimistic. There were many reasons for the speed with which the entirely new nuclear technology could be developed. I will mention only two important ones.

First, the time was technologically ripe; the discoveries of the preceding years had created a favorable technical situation. Second, Hitler's monstrous intentions were so evident that anybody, and I emphasize anybody, was willing to drop his work to strive toward Hitler's destruction. Thus, at the time of need the laboratories could assemble a gathering of young, talented people at the peak of their abilities; such a collection has never been seen before or since. This was well understood by General Groves, who once in a visit to Los Alamos, when confronted with some complaint about living conditions, smilingly said: "The government has assembled here the most expensive collection of crackpots that have ever been seen, and has put me in charge of them. I have to keep them happy." At Los Alamos, Bohr, Chadwick, Fermi, von Neumann, and Oppenheimer were the old guard; but among the young scientists were at least six future Nobel Prize winners. The average age of the scientific personnel was around thirty years.

It is often asked how far the Germans had proceeded in the development of the bomb by the end of the war, and if there was a real risk of America's being anticipated by them. We obtained reliable information on German nuclear work only after the Allied invasion of Germany. The Alsos mission, which gathered intelligence on nuclear work, found that the German project, although initiated shortly after the discovery of fission, had made relatively little headway. The Germans had not yet produced a chain reaction, nor had they prepared plutonium or separated uranium isotopes in appreciable measure. The reasons for their slowness were organizational, technical, military, and industrial. Personalities among the leaders of the effort had their weight too. Ability, rather than good will, or motivation to work, was lacking.

Information on the Soviets is scarce. A biography of I. V. Kurchatov, who held a prominent position in the Soviet work, says that although plans and talk started early, serious work commenced only in 1942, primarily at the prodding of G. N. Flerov, the discoverer of spontaneous fission; A. Ioffe, a former collaborator of Röntgen and an important personality in Russian physics; and P. Kapitza, a protégé of Rutherford who, in 1937, had been prevented by the Russian government from returning to Cambridge following a vacation in the motherland. Working conditions, however, were extremely unfavorable because of the Nazi invasion. It seems that the Soviet government gave high priority to nuclear arms only after the American explosions and the end of the war. In December 1947 the first Soviet reactor went critical; the USSR exploded its first atomic bomb in August 1949 and exploded a mixed fission-fusion bomb in August 1953. The United States had exploded a similar type of device in 1951. Without doubt the Soviets benefited from the previous U.S. experiences, which were known to them.

The English project, as mentioned before, had great importance as an encouragement to the United States and was paramount in focusing the U.S. effort and in convincing people of its purposefulness. It soon merged with the American effort. Canada also contributed importantly in the framework of the United States. Small-scale attempts were made in Japan too.

Among the various consequences of the war applications of science, some beneficial to science and some not, I must call attention to an extremely detrimental one, the introduction of secrecy. It poisoned the scientific environment and is not completely dissipated yet. Military secrecy is obviously necessary, and industrial secrecy is a well-established and justifiable practice, but secrecy in science is a self-contradictory proposition. By definition, science must impart knowledge and reveal its discoveries to all. In the history of science there have been great men—for instance, Newton—who for various reasons tried to keep their results secret, but as time went on it became apparent that this was a fatal practice, and in modern times the tendency of scientists is to publish everything as fast as possible— sometimes even too fast.

During the war military necessity forced secrecy on results of manifest and immediate military importance. We have seen an attempt at self-imposed discretion immediately following the discovery of fission. The extension of this attempt failed mainly because of the opposition by the Joliots, but it is doubtful if it could have accomplished anything. A more limited action on the part of American scientists, on the other hand, was successful.

When the Manhattan District was created, security became the responsibility of General Groves. He treated the difficult problem with good sense and flexibility. Later, however, security rules were codified and entrusted to a bureaucracy, which occasionally treated them as an end in themselves, rather than as an unpleasant necessity. There were abuses and other deleterious results. Since the war secrecy often has retarded desirable technical developments or prevented frank public discussion of important decisions without in any way enhancing military advantages. The climax is reached by persons who believe in a "secret of the atomic bomb" and similar humbug. One of the few things on which all scientists agree is the distrust of secrecy and the desire for abolishing it as soon as possible.

I have expanded on atomic energy and the Manhattan Project because of the importance of the subject, because it is very near to science, and because I have at least partial direct experience with it. One should not forget, however, that this enterprise was not unique. For instance, the history of radar has several traits in common with that of atomic energy. Here, the work and discoveries of E. Appleton, M. Tuve, G. Breit, and others who studied geophysical problems connected with Heaviside's layer, paved the way. They go back all the way to the turn of the century, when Marconi was trying to communicate at great distances and needed to overcome the curvature of the earth. The basis for military applications had been prepared, and the art had ripened at the right time to materially influence the Battle of

Britain. Special scientific military laboratories devoted to radar were created in Great Britain first, and later in the United States. Here, too, there was a spectacular concentration of talents as well as grandiose applications that followed in the postwar era ranging from astrophysics to molecular spectroscopy.

Fermi's Final Work

We return now to Fermi's life. At the end of the war he had the clear feeling that nuclear physics was reaching a stage of maturity that made it less attractive to him. It is true that with the unprecedented neutron beams obtainable from nuclear reactors he could still do experiments that were destined to open several new chapters of physics, especially in solid state, but he felt that the center of interest was shifting. With an ironical smile he quoted the Duce, Mussolini, who had said, "Either renew oneself or perish," and he was perfectly ready to renew himself.

The University of Chicago planned to build three new institutes—one cryogenic, another for the study of metals, and the third, a nuclear institute. The university succeeded in attracting several of the leaders of the Manhattan District, among them Urey and Fermi. Fermi did not want to direct the nuclear institute, preferring to entrust it to the able hands of his close friend S. K. Allison, thus freeing himself of administrative chores. He wanted to devote his whole time to teaching and research. He then founded a flourishing Chicago school that formed several of the major physicists of the postwar era, for instance, Lee and Yang. Fermi's seminars were famous for the interest of the discussions that took place on the most diverse subjects. Ideas were generated there that produced further important work. Colleagues and students could go to Fermi's office to talk to him of their own new ideas, and they often departed with new important suggestions or with a paper written in common with Fermi, who at once could develop and transform what had been only dimly perceived by the visitor.

Fermi wanted to continue experimental research in his two chief interests—high energies and applications of computers. From the Los Alamos period and his frequent exchanges with J. von Neumann he had come to appreciate the potentialities of electronic computers, and whenever he could, he returned to Los Alamos to use one of the first computers in existence. It was a primitive device compared to later models, but with its help he experimented in new directions, for instance, in statistical mechanics. In this connection I know that Fermi had invented, but of course not named, the present Monte Carlo method when he was studying the moderation of neutrons in Rome. He did not publish anything on the subject, but he used the method to solve many problems with whatever calculating facilities he had, chiefly a small mechanical adding machine.

High-energy physics experimentation was tied to a suitable accelerator. One was being built in Chicago but not as rapidly as in Berkeley, which was at that time the fountainhead of accelerators. Fermi worked directly at the building of the Chicago cyclotron, as he used to do, when he wanted something. He was a great follower of the do-it-yourself principle and applied it from mechanical work to the most long and tedious numerical calculations. While working on and waiting for the accelerator, Fermi undertook a thorough study of particle theory as known at the time. He calculated all he could, down to numbers, in order to be prepared. There is a trace of this work in the Silliman Lectures he gave at Yale University in 1951. As soon as the Chicago accelerator functioned reliably, Fermi started studying the pion-nucleon collision in collaboration with colleagues and students. He still managed to make an important discovery in this field too: the first resonance in the proton-pion collision. In the analysis of these experiments he and his associates made extensive use of the Los Alamos computer.

After the war Fermi visited Italy repeatedly and lectured there. During his last visit, in the summer of 1954, he gave a course on pion physics, but he was already suffering from an undiagnosed illness. On his return to Chicago in September an exploratory operation revealed an incurable stomach cancer. He met his end with Socratic serenity and with almost superhuman strength of character. He died on November 29, 1954, at the age of fifty-three. Thus disappeared the last physicist who dominated the whole field both in theory and in experiment. It is doubtful whether, with increasing specialization, we will ever again see such a universal excellence.

Chapter 11

E. O. Lawrence
and Particle Accelerators

Large-Scale Physics

In the last chapter I pointed out how the scale of physics experiments increased by a factor of about a million. Naturally, such an increase brings about profound changes in the nature of research. When one reduces the length scale by a million, the results are truly impressive; physics has moved from classical physics to quantum physics. An alteration by a factor of a million in the power of experimental resources also necessarily produces enormous changes. This leap in the scale of physics operations is an interesting phenomenon in itself, and I will devote some words to it.

The passage of physics to a grand scale is usually associated with particle accelerators. This is partly correct but many of the features of future developments appeared earlier: the association of science with engineering, the collective character of the work, the international status of the laboratory, the specialization of laboratories centered on one technique, the division of the personnel into permanent staff and visitors. A laboratory with all these characteristics had been formed by Heike Kamerlingh Onnes (1853–1926) at the end of the nineteenth century for the study of low-temperature phenomena.

H. Kamerlingh Onnes (Figure 11.1) was born in Groningen, the Netherlands; he first studied there and later at Heidelberg, Germany. He was influenced by the two great Dutch theoreticians of his time—H. A. Lorentz and J. D. van der Waals. Kamerlingh Onnes wrote a dissertation on Foucault's pendulum and in 1882 became professor of physics at the University of Leyden, where he remained for the rest of his life. There he devoted all his efforts to the creation of a cryogenic laboratory, which for a long time was the foremost in the world.

By working on low-temperature phenomena and on the techniques

Figure 11.1
H. Kamerlingh Onnes
(1853–1926), founder of
the Leyden Cryogenic
Laboratory, which was at the
forefront of low-
temperature physics for
many years. In this
laboratory were found many
scientific organizational
traits typical of later great
international laboratories.
(Nobel Foundation.)

for obtaining the low temperatures, Kamerlingh Onnes achieved masterpieces in both physics and engineering. The laboratory was so active that at a
certain point it needed a journal of its own for communicating its results;
this journal became a bible for low-temperature physicists. In 1904 Kamerlingh Onnes became rector of the University of Leyden, and in his inaugural
address, as well as later in his Nobel Prize speech, he clearly expressed the
goals and problems of the Leyden cryogenic laboratory. His views on the
problems presented by the laboratory and their solutions are still valid and
are applied in the great modern international laboratories. Specialized international laboratories are now devoted not only to cryogenics but also to
particle accelerators, to the generation of strong magnetic fields, to high-flux
nuclear reactors, to astronomical observation, and to other subjects.

The investigations in the Leyden laboratory comprised a great number of measurements necessary to establish cryogenic science. They included the study of the temperature scale near absolute zero, specific heats,
vapor pressures, and magnetic susceptibility; and also the development of
cryogenic machinery. The laboratory's crowning technical accomplishment
was the liquefaction of helium, achieved in 1908. The study of properties of
matter at such low temperatures revealed many remarkable phenomena, the
most outstanding being superconductivity (1911)—that is, the sudden dis-
appearance of electric resistance at sufficiently low temperature in mercury
and other substances (Figure 11.2).

Even if Kamerlingh Onnes's work is the first example of large-scale
physics, it was not particularly familiar to, nor taken as a model by, accelerator builders, who came many years after the establishment of the
Leyden laboratory.

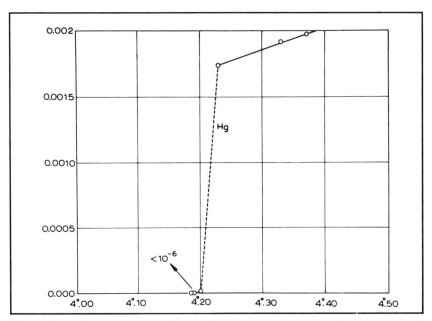

Figure 11.2 The electrical resistance of mercury as a function of temperature. At 4.2°K it suddenly vanishes, owing to the onset of superconductivity. The curve is taken from Kamerlingh Onnes's Nobel lecture given in 1913. (Nobel Foundation.)

The First Accelerators

The incentive for building accelerators arose from the need for particles more abundant and of higher energy than those obtainable from natural sources. Furthermore, particles from natural radioactive sources are practically limited to electrons, gamma rays, and alpha particles; and every indication pointed to the growing importance of protons and, after 1932, of neutrons.

After Rutherford's pioneering experiments of 1917 it became clear that the acceleration of particles with laboratory apparatus was the best way for making progress in the vitally important area of nuclear disintegration. The first attempts were made in the United States by G. Breit, M. Tuve, and others in about 1925. They built a Tesla coil (high-voltage transformer) and applied the voltage obtained to a tube suitable for accelerating particles. A little later, A. Brasch and F. Lange in Berlin used an impulse generator for accelerating protons. They even tried to obtain the high voltage from storm clouds, and it is possible that they accomplished nuclear disintegrations; the method, however, was not practical and even dangerous: A physicist lost his

Figure 11.3 John D. Cockcroft (1897–1967) at his laboratory bench at the Cavendish Laboratory in 1932. (Cavendish Laboratory.)

life, struck by lightning. C. C. Lauritsen and H. R. Crane, at the California Institute of Technology, used a series of transformers mounted in cascade, and energized with these high-tension x-ray and proton accelerating tubes. R. J. Van de Graaf at Princeton built a new type of electrostatic generator, the ancestor of a kind of apparatus still widely used.

The first disintegrations by artificially accelerated particles, however, were achieved by John D. Cockcroft (1897–1967) and Ernest T. S. Walton (b.1903) at the Cavendish Laboratory (Figures 11.3, 11.4). They developed a voltage multiplying circuit, first invented by H. Greinacher in Switzerland, and under continuous prodding from Rutherford, who wanted to see its application to physics, fed the voltage obtained to a discharge tube accelerating protons. In 1932, at 770 kV, they obtained the disintegration of lithium into two alpha particles.

In all these investigations the voltage obtained from some device was

Figure 11.4
Cockcroft and Walton's electrostatic accelerator, with Cockcroft under the discharge tube. With this apparatus Cockcroft and Walton disintegrated the nuclei of lithium and beryllium, the first to be disintegrated with artificially accelerated particles. (Cavendish Laboratory.)

applied to a discharge tube, and thus for particles with an energy of the order of 1 MeV it was necessary to have voltages of 1 MV in the laboratory. Such high voltage presented all kinds of technical difficulties and hazards. More subtle approaches that avoided high tensions were proposed as early as 1922 by various physicists and engineers; some used electromagnetic induction to accelerate the particles, others used multiple acceleration. E. O. Lawrence was foremost in bringing multiple acceleration to a practical, successful conclusion. D. Kerst developed induction accelerators (betatrons).

Lawrence and the Cyclotron

Of Norwegian descent, Ernest O. Lawrence (1901–1958) was born in Canton, North Dakota, in surroundings that still vividly reflected the American pioneer spirit with its traditional ideas: extreme respect for work, libertarian tendencies linked occasionally with fundamentally conservative or even reactionary traits, individualism, self-reliance, and provincialism. Ernest's father was a superintendent of schools, and the son had access to books, including some on science. Soon he delighted in building radio transmitters and other apparatus.

Lawrence was bright and intelligent, and after attending public schools, he completed his studies at Yale University. Despite his formal education, however, he never succeeded in acquiring a refined culture either in physics or in other fields. On the other hand, he had a real intuitive feeling for physics, a strong drive for achievement, and the ability to deal with people. At heart he remained more an inventor than a scientist. More cultivated than Edison, he had some traits in common with the great inventor. When J. J. Thomson tried to speak about x-rays with Edison, he found that it was better to change the subject; something similar happened to Fermi with Lawrence. Many physicists, even in his own Radiation Laboratory, knew nuclear physics and even accelerator science better than Lawrence. However, he was a vital element, unique in the laboratory that he created around his own invention and directed with great success in a dictatorial fashion. His extraordinary leadership, enthusiasm, and personality were more important than his science.

At Yale University Lawrence was considered a brilliant student of great promise. As a young professor he was attracted by the University of California—then in a period of expansion—and he went there in 1928. In 1929, while browsing through journals, he was struck by a drawing he saw in an article by R. Wideroe in the *Archiv fuer Elektrotechnik,* and conceived the idea of the cyclotron (Figure 11.5).

As we have already seen, an ion of specific charge e/m moving in a uniform magnetic field with an initial velocity perpendicular to the lines of force describes a circle of radius $r = mcv/eB$ with an angular velocity $\omega = eB/mc$ independent of v and r. If we place an ion source in a magnetic field and add an alternating electric field perpendicular to B, on a diameter of the trajectories and varying with a frequency $\nu = \omega/2\pi$ in such a way as to accelerate the ions whenever they cross the diameter, we achieve multiple acceleration. On each crossing of the electric field the ions acquire a certain energy, and this energy is multiplied by the number of crossings. The electric field with the required characteristics is obtained by the use of two boxes in the form of the letter D—hence called *dees*—facing each other. The dees are kept at alternating voltages varying with a frequency ν. The whole system is contained in a vacuum chamber (Figure 11.6).

Each time an ion passes through the gap separating the dees, it receives the energy eV, where V is the potential difference between the dees. The trajectory of the ions is a spiral originating from the source, and at the end the particles are deflected into a channel through which they escape

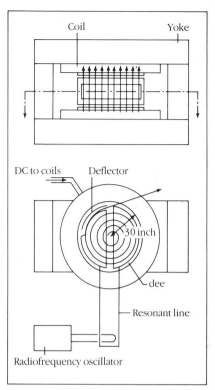

Figure 11.5 Ernest O. Lawrence (1901–1958). In his hand is one of the first cyclotrons, of minimal dimensions compared to the successive ones. (Lawrence Berkeley Laboratory.)

Figure 11.6 A schematic drawing of the principle of the cyclotron. [From E. Segrè, *Nuclei and Particles,* 2nd ed., 1977, with permission of Benjamin/ Cummings, Inc.)

from the evacuated box. Because of the multiple accelerations the potential differences existing in the instrument are all very small compared with what would be required to obtain the same energy in a single acceleration. Thus the difficulties inherent in high voltages are avoided, a technical advantage of inestimable value.

Perhaps the idea of the cyclotron was not entirely new when it occurred to Lawrence, but new were the vigor and purposefulness with which Lawrence undertook its practical realization, with the help of a series of co-workers starting with N. E. Edlefson and M. S. Livingston. For the remainder of his life Lawrence built a series of cyclotrons or other accelerators of increasing energy. This progression recalls that of G. Marconi trying to communicate over increasing distances.

Lawrence needed funds and technical help. He turned to private individuals or foundations for funds, stressing, at least in the early years, the possible medical application of the cyclotron, an aspect that was also very dear to his humanitarian aspirations. For recruiting technical assistance, he was helped by his contagious enthusiasm, the manifest success of the enterprise, and the unemployment conditions of the 1930s among physicists. Lawrence himself was most generous in allowing outsiders to use the machine, in supplying radioactive substances produced by the cyclotron to fellow scientists nearby or far away, and in refusing to take credit for work to which he had indirectly but vitally contributed by furnishing indispensable technical means. He was also ready to supply blueprints and help to anybody who wanted to build a cyclotron.

He soon realized that it was easier to find technical than leadership ability, so he concentrated his great administrative talents on organization in a broad sense and left the technical work almost entirely to his co-workers. Taking advantage of the attraction that his laboratory exerted on young talented physicists, he was able to recruit excellent scientific personnel. He observed them closely and, with cool judgment, he let them use their particular aptitudes to the general benefit of the laboratory. In this, he left them considerable freedom of action but required unlimited devotion to work. He was always very optimistic in technical matters and believed that many accelerator problems could be solved by brute force, or, alternatively, that persistence would uncover solutions to difficulties. Once Lawrence suggested a certain modification of the pole pieces of the cyclotron. The laborious adjustment was made, with a negative result. His typical comment was "Excellent! If it spoils the beam, we will try the opposite change and it will improve it."

Over a period of time he assembled a group of first-class scientists who became experts on accelerators, among them M. S. Livingston, E. McMillan, L. Alvarez, R. Thornton, D. Cooksey, W. Brobeck, and R. R. Wilson. Indeed, a large fraction of the first generation of accelerator builders was formed in Berkeley, at the Radiation Laboratory (now Lawrence Berkeley Laboratory). Don Cooksey (1892–1977) had a special place among them. He was a gentleman of means who had a Yale Ph.D. in physics. He recognized Lawrence's ability very early and dedicated his life to helping him. He admired, understood, and liked him greatly, and his help extended to all fields—technical, financial, human. His devotion and loyalty were unlimited and completely unselfish and unobtrusive. He had a vital part in the success of the Lawrence enterprises.

The first cyclotron had a diameter of a few inches. Its glass vacuum chamber, in which ions were supposed to circulate, could be held in one hand. The next one was a little larger, and in 1930 successful operation was achieved. The first disintegrations obtained with a cyclotron were reported in 1932. When I first arrived in Berkeley in 1936, there was already a 37-inch cyclotron, built with a surplus magnet from the Federal Telegraph Company, that produced strong sources of radioactive isotopes (Figure

11.7). A 60-inch cyclotron followed in 1939 (Figure 11.8). Every one of these machines was used for important physics or chemistry investigations.

In the early 1930s Lawrence tried his hand at nuclear physics with the cyclotron. Some of the results he obtained were in error, and he had protracted correspondence on the subject with the physicists of the Cavendish Laboratory, who did not agree with him. This episode may have contributed to turning him almost exclusively to machine development, a happy decision for scientific progress because this was the field where he really excelled. In any case, he became primarily interested in the performance of the cyclotron, paying relatively little attention to its application to science. A year or two before Curie and Joliot discovered artificial radioactivity, Lawrence was making in the cyclotron radioactive substances in amounts thousands of times larger than those observed by the Curie-Joliot. Yet he did not notice that he was surrounded by radioactivity. Later the same situation arose with fission. Although interested in scientific discoveries, Lawrence was not ready to sacrifice machine development to nuclear research. Most of the machine time was devoted to maintenance, improvement of the machines, or medical research. This precluded the possibility of concentrating technical and intellectual effort on the exploitation of the machines in the nuclear field.

Lawrence's unflagging devotion to machine development was not always appreciated by the young physicists surrounding him. Whenever there was an opportunity, they strove to use the cyclotron, almost surreptitiously, for scientific purposes.

In retrospect, I have come to think that Lawrence's singleness of purpose was necessary for the successful development of the machines. In other laboratories physicists built cyclotrons and tried to use them as soon as possible for scientific investigations, with rather unsatisfactory results because the machines were too often under repair and could not be used for sustained work. It was also difficult to find personnel skilled in both fields, and only much bigger laboratories than those existing and endowed with much greater resources could have sustained both purposes.

The Radiation Laboratory was organized very differently from the usual university laboratories, and for this reason the Regents of the university recognized it as an entity separate from the Physics Department as early as 1932. In spite of some inevitable jealousy and friction, a peaceful and even cordial coexistence between the Radiation Laboratory and the Department of Physics prevailed, with inestimable advantage for both.

The financing of the Radiation Laboratory was primarily Lawrence's responsibility, and it was a difficult problem for him. Money was very scarce and the economy depressed. Lawrence had to spend much time in seeking possible donors and in promotional activities, mostly away from Berkeley. He was unusually good at public relations work and probably found it congenial. Benefactors were impressed by his personality, op-

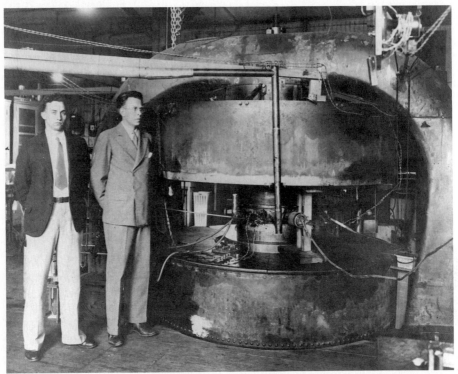

Figure 11.7 M. S. Livingston and E. O. Lawrence in LeConte Hall at the University of California, Berkeley, standing next to the 37-inch cyclotron. Originally it measured 27 inches; it was increased to 37 inches in 1936. It was used to measure the magnetic moment of the neutron and to produce the first artificial element, technetium. (Lawrence Berkeley Laboratory.)

timism, and success, and by the responsible way in which he fulfilled his claims. The laboratory was run on a tight economy, husbanding money whenever possible.

Policies and Personalities

In the 1930s J. R. Oppenheimer, professor of theoretical physics at Berkeley, was Lawrence's intimate friend and consultant, although they were almost opposites in science and politics. The genuine close friendship that existed between such different personalities may be surprising; the uncovering of its foundation would require a careful analysis of the two people. Perhaps they admired in each other their complementary qualities.

Later, when the European situation was moving precipitously toward

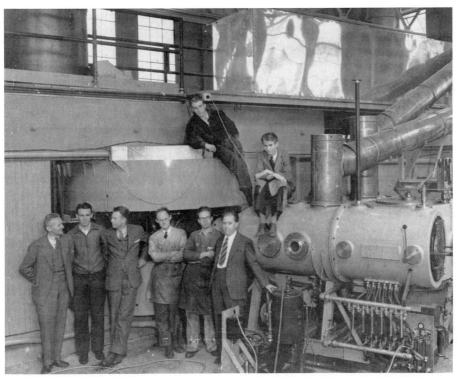

Figure 11.8 The 60-inch cyclotron at the Crocker Laboratory at the University of California, Berkeley, in 1944. Standing, from the left: D. C. Cooksey, D. Corson, E. O. Lawrence, R. L. Thornton, J. Backus, W. Salisbury; on the cyclotron are L. W. Alvarez and E. McMillan. Many transuranic elements were produced with this cyclotron. (Lawrence Berkeley Laboratory.)

catastrophe, Oppenheimer echoed the slogans of the extreme left and Lawrence, true to his Midwestern traditions, was rather isolationist. Furthermore, he did not grasp the possible importance of atomic energy for war purposes. However, when war broke out in Europe, Lawrence felt that the survival of Great Britain was essential and that one had to help in every possible way. He persuaded his closest friends and associates, McMillan and Alvarez, to help on the project of most immediate use, radar work at Cambridge, Massachusetts. Atomic energy and weapons looked like science fiction to Lawrence, and he was not alone in that opinion. In fact, many leading American physicists thought similarly, or at least they gave a much lower priority to atomic programs than to radar work, justifiably, perhaps, in the immediate perspective. For a certain period nuclear energy was advocated primarily by European physicists who had come to the United States, fleeing from Hitler and Mussolini. Later, Lawrence, Compton, and other major American figures grasped the potentials of nuclear work and took positions of leadership. They soon understood that the dimensions of the enterprise imagined by the Europeans were totally unrealistic. The Americans erred by underestimating the effort needed by perhaps a factor of ten, but the Europeans were off by a factor of a thousand.

After the Pearl Harbor attack, Lawrence's attitude changed radically, and he immediately put the laboratory on a war footing. Having learned that uranium isotope separation was of paramount importance, he decided to attack this difficult and fundamental problem in the most direct way, by building hundreds of gigantic mass spectrographs. He persisted in this particular way of separating isotopes and succeeded in the end. The expense was tremendous, and success depended, at least in part, on the parallel development of diffusion methods that enabled the spectrographs to be fed with slightly enriched material. One of the first atom bombs, dropped on Hiroshima, actually contained U^{235} separated at Oak Ridge by Calutrons, the name given to Lawrence's mass spectrographs.

During the war Lawrence's friend Oppenheimer became director of the Los Alamos laboratory, where the atomic bomb was made, and thus had great power and responsibility for the first time in his life. At the end of the war both Lawrence and Oppenheimer had matured considerably, and especially Oppenheimer had changed his naive leftist views. Both were much in the public eye, a completely new experience for Oppenheimer. They drifted apart and ended as standardbearers of opposite factions in a bitter controversy concerning the building of a hydrogen superbomb. Several authors have written books and dramas on their fights. Often these literary outpourings are extremely partisan, mostly against Lawrence; on the other hand, there is an official biography of Lawrence that resembles hagiography.

The roots of the antagonism between Lawrence and Oppenheimer lie in their personalities and in the new circumstances in which they found themselves. They were now able to impress their different views on the universities with which they were associated, on science policy, and even on national policy. They were each backed by enthusiastically loyal close friends, who reflected their views. Most important, Lawrence and Oppenheimer made very different appraisals of the international situation and of what was desirable and possible for the United States and the world. Oppenheimer had an internationalist point of view and a realistic view of the power and limitations of the United States, and he feared future conflicts with Russia. Lawrence was more confident in U.S. power and believed more in forceful technical solutions and the effectiveness of armaments. Both were deeply convinced of the righteousness of their causes and believed they were working for peace. Obviously their differences are still alive.

Following the first nuclear explosion by the USSR (August 29, 1949), Edward Teller (b.1908), a theoretical physicist of Hungarian origin, Lawrence, and L. W. Alvarez convinced themselves that the safety of the United States required an all-out effort toward building a thermonuclear bomb of unlimited power—that is, power determined only by the amount of material used. The government agencies charged with advising on the subject, in particular the General Advisory Committee of the Atomic Energy Commission (J. R. Conant, H. Rowe, C. S. Smith, L. A. Dubridge, O. E. Buckley, Oppenheimer, Fermi, and I. I. Rabi), were of a different opinion, which was also shared by the majority of the Atomic Energy Commission. Teller, Lawrence, and their partisans then launched a campaign to persuade political and military personalities to their views. Many parties were involved, including high-ranking officials of the armed forces, congressmen, senators, and cabinet members. Secrecy prevented public discussion and poisoned the atmosphere. President Truman made the final decision in January 1950, favoring the activists.

Some of the programs promoted by Lawrence were technically ill-conceived and failed, with a great waste of money. However, the hydrogen bomb could be built because of an invention, made at Los Alamos, by the Polish-born mathematician S. Ulam and E. Teller. That laboratory then built the bomb, which was exploded in November 1952.

Rivalries within the laboratory and personality conflicts brought about a rift between Teller and the director of the Los Alamos laboratory, N. Bradbury, who had succeeded Oppenheimer in that post. As a consequence Lawrence campaigned for the construction of a second weapons laboratory at Livermore, California, over which he would have control. This laboratory was started in 1952, but the animosities between the principals in these quarrels continued and contributed to the Oppenheimer security clearance revocation mentioned on p. 216.

The Livermore Laboratory also became a source of leadership for the U.S. military establishment. H. F. York, H. Brown, and J. Foster, all young physicists attracted by Lawrence, in succession directed the Livermore Laboratory and later guided the American military research and development program for a long period.

In the last years of his life Lawrence invented and developed a color television system; however, it had little impact on the industry. Lawrence, no doubt, desired peace sincerely. Unfortunately, his views were often naive and ultimately contributed to the armament race that has diminished everybody's security. While he was attending, as an expert, a Geneva disarmament conference, he was stricken by a recurrence of the colitis that had affected him for a long time. He rushed home, but a few weeks later on August 27, 1958, he died following surgery.

Racing for Ever-Higher Energies

At the end of the war there was a period of indecision regarding the support of scientific research in the United States. The National Science Foundation had been proposed but not yet established. The navy, through its Office of Naval Research, was doing an excellent job in a field remote from its traditional activities. The newly created Atomic Energy Commission (AEC), succeeding the wartime Manhattan District, was finding its bearings. Lawrence had a clear vision of what he wanted to do and good relations with General L. R. Groves, based on his war performance and personal trust. He succeeded in obtaining early and unflagging support, first from the general and later from the Atomic Energy Commission. However, he preserved a very large degree of autonomy under a contract between AEC and the University of California. The penurious conditions of the prewar era were thus replaced by generous support, and Lawrence could return under favorable circumstances to his favorite endeavors. Before the war Lawrence had planned to build a cyclotron of the then-immense energy of 100 MeV. He expected to overcome the relativistic difficulties by a method of brute force.

The magnet for the machine had already been built, and during the war it had been converted to a magnet for studying Calutrons. I do not know whether the machine as originally conceived would have worked satisfactorily, but fortunately it was not necessary to test it: During the war McMillan, at Los Alamos, invented a method for bypassing the relativistic difficulties. The method uses the principle of *phase stability.* Strangely, the same method had been invented a little earlier, unbeknown to McMillan, by V. I. Veksler in the Soviet Union. The two, cool heads and gentlemen, soon recognized what had happened. Aware of their mutual good faith, they avoided a priority dispute and became and remained good friends until Veksler's untimely death.

A few years later, a second invention was also necessay to overcome the difficulties inherent in building even higher energy accelerators. This is the method of *strong focusing.* It also has a curious history. In 1949 a Greek-American engineer, N. Christofilos, utterly unknown in the scientific world, sent Lawrence a patent application describing "strong focusing." Particle beams in accelerators must stay together during a huge number of revolutions in the machine, permitting the repetition of the acceleration many times, a feature necessary for reaching high energies. Strong focusing permits containing the orbits in relatively small pipes, a result of great practical importance. Lawrence probably did not read the patent application. He passed it to a colleague, who, after cursory examination, failed to understand it.

A few years later, in 1952, E. D. Courant, M. S. Livingston, and H. Snyder, without any knowledge of Christofilos's work, reinvented the same method, described it in an elegant mathematical form, and planned an accelerator based on it. When Christofilos learned of this, he was justifiably upset. It was an embarrassing situation for the AEC. The government agency resolved the problem by recognizing the facts and hiring Christofilos in one of its laboratories, where he continued making various inventions until his sudden death in 1972 at the age of 55.

Figure 11.9 The 184-inch synchrocyclotron built after the war. Some of the numerous personnel who worked on the construction of the machine appear in the photograph. The first artificial mesons were produced with this machine. (Lawrence Berkeley Laboratory.)

The immediate postwar era saw a true jump in the dimensions of accelerators. At Berkeley the scientists and engineers of the Radiation Laboratory, using the magnet mentioned earlier, built a 184-inch synchrocyclotron (Figure 11.9) and an electron accelerator, the synchrotron, incorporating the phase-stability principle. The Berkeley synchrocyclotron was the machine that produced artificial mesons for the first time. It was followed by the Bevatron, which reached 6.4 GeV and created proton-antiproton pairs (Figure 11.10). Similar apparatus were also built in other parts of the world. At the University of Illinois in Urbana, for instance, D. Kerst built a large electron accelerator in which the acceleration was obtained by induction, without any external electric fields. This type of accelerator, called betatron, had been invented and tested by Kerst on a small scale before the war. In Chicago Fermi and his colleagues built a synchrocyclotron similar to the one in Berkeley; at Brookhaven National Laboratory, a new laboratory built on Long Island, New York, by a consortium of Eastern universities, M.

Figure 11.10 The Bevatron during its construction. It is a synchrotron for protons, or a protosynchrotron, which reached the energy of 6.4 GeV and produced the first proton and antiproton pairs. Note the size of the man in the bottom right-hand corner. (Lawrence Berkeley Laboratory.)

Livingston and others built a cosmotron, which reached 3 GeV. The Soviets secretly built a 10 GeV accelerator at Dubna and revealed its existence at an Atoms for Peace Conference in Geneva in 1955.

With the changes in organization, financing, and structure of these enterprises, science came to depend increasingly on national government rather than local or private support. As a result national governments came to have a much more direct influence on science policy, and this brought about the formation of national laboratories. Recently, the support basis has been extended to many nations, and international laboratories have developed. There is even talk of a world accelerator that would reach an energy of 10^{13} eV; it would be supported by willing nations at the cost of 1 billion dollars.

The prototype of the international laboratories is CERN (Centre Européen pour la Recherche Nucléaire) in Geneva, Switzerland, which is

Figure 11.11 An aerial view of the Fermi National Accelerator Laboratory, Batavia, Illinois. The largest circle is the main accelerator; the radius is 1 kilometer. Three experimental lines extend at a tangent from the accelerator. The 16-story twin-towered central laboratory is seen at the base of the experimental lines. (Fermi National Accelerator Laboratory.)

supported by many European nations. It helped bring Europe to a status comparable with the United States in particle physics, a position lost before and during the war due to the follies of the European governments and the war catastrophes.

Today the largest accelerators are at the Fermi National Accelerator Laboratory (FNAL) at Batavia, Illinois, which maintains the energy record (500 GeV); at Serpukhov in the USSR (76 GeV); at Brookhaven National Laboratory (33 GeV); and at CERN (300 GeV), where, however, there are storage rings that allow even greater center of mass energies than at FNAL, albeit with weak intensity. CERN also has a 500 GeV accelerator that was completed in 1977 (Figure 11.11).

We have talked mainly of proton accelerators. Electron acceleration proceeded on lines more or less parallel to that of heavy particles. Relativistic mechanics was absolutely necessary for its development just as Newtonian mechanics was essential for going to the moon; this is one of the most direct proofs of relativity, if there were still any doubts. For dealing with energy losses due to radiation that arises when the fast-moving electrons travel in a circle, linear accelerators are preferable to circular ones. For very high energies one returns to linear accelerators such as the 2-mile-long Stanford linear accelerator (SLAC) in Palo Alto, California, which produces electrons, positrons, and photons. Combined with storage rings and colliding beams, it reaches center of mass energies of 8 GeV.

The race for high energies is beset by a serious problem when the energy of the particle is large compared with the rest energy of the projectile and of the target. The important energy is not the laboratory energy of the projectile, but the center of mass energy of the projectile and target. It is the latter that is available, for instance, for creating new particles. For a proton hitting a proton, this center of mass energy is given in the extreme relativistic case by the expression $E = \sqrt{2Mc^2 E_{\text{lab}}}$. Mc^2 for a proton is 0.938 GeV, and thus a proton of 1,000 GeV in the laboratory hitting another proton at rest has an energy in the center of mass of only 43 GeV! The cost of the machines increases proportionally with the laboratory energy, or perhaps even faster; it is thus apparent that racing for high energies by increasing the laboratory energy is financially a losing proposition.

By colliding two beams of 22 GeV protons, one obtains in the center of mass the same energy as with a 1,000 GeV accelerator, although the intensities attainable are much smaller. Colliding beams seem at present the only answer for extremely high center of mass energies.

In 1979 the record for center of mass energy for electron positron collisions is around 30 GeV, obtained at Desy (Deutsches Elektronen Synchrotron) near Hamburg, Germany. Cosmic rays, however, maintain the absolute energy record: 10^{20} eV. Such energies are unattainable by any laboratory means for the foreseeable future, but cosmic rays are uncontrollable and, at extremely high energy, very rare. It is difficult to use them for particle research.

The need for machines has conditioned the progress of particle physics. This forefront field of research will be treated in the next chapter.

Chapter 12

Beyond the Nucleus

At the end of the war physicists committed to war work had to consider their future. The great majority returned to the universities they had temporarily left to join the war effort, some as professors and some as students. Meanwhile, the center of gravity of physics had shifted from Europe to the United States, partly because Europe had suffered great damage from bombing, whereas the United States had been spared, and partly because the great war enterprises such as radar, the atomic bomb, and the beginning of computers had established a scientific supremacy in the United States. Not least, the foolish policies of the axis countries had deprived them of a considerable part of their scientific manpower. Furthermore, living conditions in a devastated Europe were unattractive even after the war.

In physics there were some obvious big changes: In nuclear physics the amount of experimental data had increased immensely, and it was clear that further increases, both qualitative and quantitative, were in the offing. New inventions in accelerating machines and nuclear reactors provided sources of radiations of unexpected power. Other techniques derived from radar contributed to opening entirely new possibilities.

Science had captured the public imagination, and consequently generous financing was available from public funds. In the United States the National Science Foundation had been proposed to finance scientific research, but the legislative process for implementing it was slow and relatively ineffectual. In the meantime, the navy and the Atomic Energy Commission, a new creation of the postwar years that should have controlled all activities connected with nuclear energy, decided to intervene, funding pure research until the situation was clarified. The action of the navy is remarkable from a historical standpoint. The high command decided that, for its own purposes, it was more useful to finance a vigorous but completely free science program without any explicit or implicit strings attached than to add one or more big ships to the fleet. Thus the navy financed all kind of scientific research, of excellent quality, but with no obvious connection with naval purposes.

The students of the postwar era were also very different from the previous generation; they were more mature and often technically better prepared. A soldier who had worked as a technician at Oak Ridge had acquired practical knowledge equivalent to several years of university laboratory training, and a sailor who had served in the U.S. Navy frequently had a good background in electronics.

All this contributed to the great vigor of scientific enterprises during the immediate postwar era in the United States.

The Elementary Particles

A new field was about to emerge, or, rather, an old field was about to blossom: the study of the so-called elementary particles. Before the war protons, neutrons, electrons, positrons, photons, and neutrinos were familiar; and in cosmic rays particles of about 200 times the mass of the electron, then called mesotrons, had been seen. The components of the world looked relatively simple, even if not as simple as in 1929 when only protons, electrons, and photons were known and seemed sufficient to explain the constitution of matter. It was also clear to many of the best informed physicists that elementary particles would be the next frontier in physics. They presented the most fundamental problems, and new experimental tools, in the offing, would allow advances in uncharted territory.

Other fields, such as nuclear physics and solid-state physics, were also full of promise. However, the problems they presented, although interesting and more important for applications than those of particle physics, were less fundamental; that is, less general, less new, and less apt to change some established concepts or to introduce some new and unexpected ones.

Other sciences connected to physics but somewhat removed from its mainstream, such as geophysics, astrophysics, and molecular biology, also looked very promising. For a physicist, however, they required a retraining in completely new fields and the development of new mental habits that are not easy to acquire. A few physicists made the effort necessary for the transition, sometimes with handsome rewards.

Physicists who wanted to be in the forefront of their own science would veer toward particle physics.

At the end of the war cosmic rays were still the only source of high-energy particles. One could observe them only when God sent them, and there was no possibility of experimenting by changing their energy or creating varied conditions, as is customary in other fields of physics. On the other hand, considerable progress had been made in detection methods, and for a few years cosmic rays competed with accelerators as a source of particles, especially because cosmic rays had energies higher than anything attainable in the laboratory at that time. Cosmic rays also had the distinct advantage of costing nothing, and especially for the European nations, devastated by the war, this was a godsend. In particular, the English and Italian tradition of cosmic ray studies was revived immediately after the war.

Efforts in the United States centered on achieving cosmic ray energies by developing bigger and better accelerators; in Europe, where this was at least temporarily financially impossible, efforts focused on improving detection methods. The most powerful and also one of the simplest advances was the improvement of photographic emulsions that made detectable singly charged particles moving with a velocity near that of light. Slower particles ionize more heavily and are thus easier to detect, but particles moving with a velocity near that of light are of paramount importance for meaningful analysis of the events observed. English and Italian investigators using cosmic rays and photographic emulsions were able to make truly important discoveries; they had thus found a cheap and extremely interesting field of physics. Later, of course, it was found that the cheapness of particle physics was illusory—in fact, it turned out to be one of the most expensive fields of physics—but by then financial conditions had improved, and Europe, too, could afford giant accelerators.

The New Science in Japan

Let us now turn to the physics. For the first time we see a Japanese physicist, Hideki Yukawa (1907–   ) preeminent in new theoretical ideas.

Before speaking about Yukawa I will digress slightly to consider the development of Western physics in Japan. At the time of the Meiji restoration in 1868 Japan had a reasonably advanced indigenous technology, but no science in the Western sense. When Japan's leaders decided to modernize the country, following Commodore Perry's visit in 1853 and the forced opening of harbors to foreign trade, they perceived the importance of "science" and imported a French and an English professor of physics and charged them with teaching their discipline in Tokyo. Linguistic difficulties were such that the two schools remained separate, and for a time there existed an English physics and a French physics. Some samurais were also dispatched to Europe to learn physics. Apparently their scientific interest was very limited, but they were full of patriotic ardor and they studied science with a military spirit. An autobiography (available only in Japanese) of one of these samurai shows him to be violently anti-Western and full of contempt for the Occidental barbarians, but these feelings are mixed with admiration for Western successes, in retrospect certainly a dangerous combination. This samurai was initially skeptical about the capabilities of Western science, but when the captain of the ship on which he was traveling announced that they would meet another ship in the middle of the ocean that afternoon, and the event happened, the samurai conceded.

After these early Japanese pioneers, men who were better prepared followed. They had some understanding and interest in science per se and were not merely performing a patriotic service. Among them Hantaro Nagaoka (1865–1950) is the most important figure. He visited Europe at the turn of the century and seemed informed about the currently important physics problems. Nagaoka became one of the leaders of Japanese science, also exercising great influence as an administrator. He taught physics for many years at Tokyo University, but when in 1931 a new Imperial University was founded at Osaka, he became its president, although without enthusiasm.

Science and patriotism were the dominating passions of Nagaoka. A letter, remarkably written in English, in 1888 to his professor Tanakadate who was temporarily in Glasgow, Scotland, eloquently shows Nagaoka's feelings, although they may have changed with the passing of time. He writes

> We must work actively with an open eye, keen sense, and ready understanding, indefatigably and not a moment stopping. We must not allow those classes of people within our doors and interrupt our work, who though appearing as if they were intently at work, soon stop working whenever there comes anything that attract[s] the eye, or the mouth, or the purse. There is no reason why the whites shall be so supreme in everything, and as you say, I hope we shall be able to beat those *yattya hottya* [pompous] people in the course of 10 or 20 years: I think there is no use of observing the victory of our descendants over the whites with the telescope from *jigoku* [hell]. Another great requisite in beating those whites is how to make our work known. This is a great difficulty. As a first step we can not write in Japanese and make the westerners understand our writings. We must borrow their language and make the whites understand. Indeed, as you admit, the whites can speak upon anything, but in our case, we are sometimes unable to speak even though there is sufficient material for talk. I think this is our great defect, and we must, if possible, learn to write and speak clearly and fluently. I don't think there is any choice of language, be it English, French, or German. Please reflect on this point.
>
> Now quitting these dreams, I shall speak of something that is going on here. ... [K. Koyzumi, "Historical Studies in the Physical Sciences," vol. 6, p. 87 (1975)]

Nagaoka published many papers in Western journals—on magnetism, spectroscopy, and other subjects. Most notable is a 1903 proposal of an atomic model similar to the solar system. Magnetism has traditionally been a preferred subject by Japanese physicists and one to which they have contributed importantly both in theory and experiment.

Scientifically, however, the most important Japanese influence was exerted by Y. Nishina (1890–1951). He first studied as an electrical engineer in Tokyo, but later went to Europe and turned to theoretical physics. In Copenhagen he worked in Bohr's institute, where, with the Swede O.

Klein, he calculated the cross-section for Compton scattering, at that time a tour de force in theoretical technique. He also worked with Heisenberg in Germany. Finally he returned to Japan, where he joined the Institute of Physical and Chemical Research, financed in part by private funds and in part by the government. This institute exerted a major influence on Japanese science. A large proportion of the most distinguished younger Japanese physicists are directly or indirectly connected with the Nishina school.

Most important because of his own discoveries and as a teacher is Hideki Yukawa (Figure 12.1), born in Tokyo in 1907, the son of a geology professor Takuji Ogawa, but in 1932 he was adopted in his wife's family, by name Yukawa. His education took place entirely in Japan, mostly in Kyoto. To a certain extent he was self-taught. With his friend and schoolmate Sin-itiro Tomonaga (1906–1979) he learned quantum mechanics partly from the original papers and partly from books. On Nishina's return from Europe, Tomonaga, who was to become famous for his studies on electrodynamics, went to the Institute of Physical and Chemical Research. Yukawa obtained a position at Osaka University, where there was a physicist educated in Europe, Seishi Kikuchi, who had transferred from the Institute of Physical and Chemical Research.

In 1935 at Osaka, Yukawa wrote an epochal paper. The paper was incomplete and perhaps not entirely correct, but it contained fundamental new ideas, which turned out to be extremely productive and have deep influence on subsequent developments.

Yukawa formed a flourishing school of theoretical physicists at the University of Kyoto. He founded also a journal *The Progress of Theoretical Physics,* which acquired world importance, especially in theoretical particle physics. Often ideas appeared in it independently of Western literature and were later or simultaneously discovered by authors that had not had access to the Japanese journal. This happened especially during and soon after the Second World War.

The mathematical language and the English idiom form a bridge between Japanese and Western physics, but occasionally one feels the different cultural basis. Of course, this may be an advantage for the progress of science, because it gives widely separated points of attack.

Yukawa was the first Japanese to win the Nobel Prize in Physics, in 1949, and this gave him enormous prestige in his country. The Japanese came to see him as living proof of having reached scientific parity with the West. Thus Nagaoka could still see his dream fulfilled before dying in 1950.

Here is a simple version of Yukawa's arguments. It is known that photons are associated with the electromagnetic field and its forces. Yukawa asked what would be the characteristics of the quanta associated with the field of nuclear forces. With simple reasoning involving little more than an application of the uncertainty principle and of relativity, he reached the startling conclusion that the quantum of nuclear forces must have a finite rest mass. He estimated that this rest mass should be about 200 times as

Figure 12.1 Hideki Yukawa (third from left) on a visit to Berkeley in 1948. Also pictured, from the left, are Fermi, Segrè, and G. C. Wick. Yukawa was the first Japanese physicist to receive his scientific education entirely in Japan. His theory interpreted mesons as quanta of nuclear forces. (Photo E. Segrè.)

large as the electron mass, or that the energy mc^2 of this quantum, when at rest, should amount to about 100 MeV. Furthermore, this particle should appear in three forms: electrically neutral, positively charged, or negatively charged with charge equal in magnitude to that of the proton.

Yukawa's reasoning was as follows: Nuclear forces have a sort of sphere of action of 10^{-12} to 10^{-13} cm. This means that at this distance, r_0, the force decreases suddenly. By contrast, in the potential of ordinary electric force $U(r) = e/r$ there is no such characteristic length. Now it is possible to interpret the force between two nucleons as due to the emission of a quantum by one of them and the absorption of the same quantum by the other. The quantum needs a time greater than r/c to go from the emitting to the absorbing nucleon, because, due to relativity, it cannot exceed the velocity of light c. During the travel time conservation of energy is violated because one has the additional mass of the quantum and the corresponding energy mc^2. However, one cannot measure energy with a precision greater than

$$\Delta E = \frac{h}{2\pi} t$$

where t is the time available for the energy measurement. This is because of the uncertainty principle of quantum mechanics. We can thus "borrow" energy in the amount ΔE, and there is no way of ascertaining a violation of the conservation of energy. If we now identify ΔE with mc^2 and t with r_0/c, we find

$$m = \frac{h}{2\pi r_0 c}$$

The assumption of a finite mass for the quanta automatically limits the radius of action. Introducing 2×10^{-13} cm as a value for r_0 and the proper values for the universal constants, we find for m about 200 times the electronic mass, or, for mc^2, an energy of about 102 MeV.

Empirical evidence shows that the forces between neutron and proton, proton and proton, and neutron and neutron are equal, and these require charge $\pm e$ or zero for the quanta. Thus Yukawa anticipated the existence of particles of a mass about 200 times that of the electron, either neutral or with one positive or negative unit of charge, and which would interact strongly with nuclei.

Fermi had had similar ideas when he developed the theory of beta decay. He thought that neutrinos could be quanta of some field, but he did not publish anything on the subject because he could not obtain significant results. Yukawa had studied Fermi's paper on beta decay and he explicitly quoted it. Yukawa's paper appeared in 1935 in the *Journal of the Mathematical and Physical Society of Japan;* it was noted but did not cause a great stir. At the time nobody had yet seen particles similar to those postulated by Yukawa, and thus they appeared mainly as an interesting speculation.

Discovery of the Pion

Only in 1937 did investigators of cosmic rays such as C. D. Anderson (the discoverer of the positron), with his collaborator S. H. Neddermeyer (who later made important inventions used in the first atomic bomb), M. L. Stevenson, J. C. Street, R. B. Brode and others begin to find in cosmic rays particles of an intermediate mass between that of the electron and the proton. The best measurements gave masses about 200 times that of the electron. These particles, called *mesotrons,* are unstable and, when free, decay with a mean life of about 2 microseconds. The mean life was inferred by subtle reasoning, based on observations of cosmic ray intensity at different elevations and at different angles above the horizon, and later directly measured by F. Rasetti. However, the cosmic ray experimenters at the time of their first observations were not aware of Yukawa's work. The war slowed down the experimental work and isolated Japan from the West. Japanese physicists had been impressed by the existence of particles of a mass comparable to that postulated by Yukawa. However, they had also noted difficulties that prevented the identification of mesotrons with the Yukawa particles; above all, the mean life of mesotrons was much too long. Furthermore, when mesotrons stop in matter, they usually, though not always, interact with nuclei of the stopping medium. The study of this phenomenon yielded an important experimental discovery by the three young Italian physicists: M. Conversi, E. Pancini, and O. Piccioni.

These three men were hiding from the Germans who would have deported them to forced labor in Germany, and they were working clandestinely in a cellar in Rome. They found that positive and negative mesotrons stopped in matter behave differently. The positive mesotrons decay more or less as though in a vacuum. The negative ones, if stopped by heavy nuclei, are captured and produce disintegrations, but when captured by light nuclei such as carbon, a good fraction of them decays, as though in a vacuum. This was not the behavior expected of Yukawa's particles. They should have reacted violently with light or heavy nuclei because the specific nuclear forces should have produced disintegrations as soon as the mesotron came sufficiently near a nucleus. The experiment showed that this was not the case, and hence mesotrons could not be identified with Yukawa's particles.

The situation was odd indeed. Yukawa had predicted particles of about 300 electron masses; one had found them but they were not the particles predicted. Theoretical physicists puzzled over the results of Conversi, Pancini, and Piccioni, which, however, seemed solid from the experimental point of view. The theoreticians endeavored to find an explanation. Tanikawa, Sakata, and Inoue in Japan as well as H. A. Bethe and R. Marshak in the United States independently put forward a hypothesis that would have removed the difficulties. They proposed that the observed mesotrons were the decay product of Yukawa's mesons, which nobody had yet observed. Making an attractive and plausible hypothesis is one thing, but ascertaining a fact is very different.

At this point a new experimental technique, or, rather, the improvement of an old one, provided a powerful tool. Before the First World War

Kinoshita, a Japanese physicist in Rutherford's laboratory, had shown that alpha particles crossing a photographic emulsion leave on their track a collection of developable emulsion grains. Thus one can see the trajectory of the particle. (We may ask, what about quantum mechanics? the uncertainty principle? the wave nature of the particles? The reader may rest assured that there is a satisfactory answer to these questions, given in detail, for instance, by Heisenberg.) The emulsions used by Kinoshita were sensitive only to relatively heavily ionizing particles; electrons were undetectable.

Professor Cecil Powell (1903–1969) of Bristol, a former pupil of Rutherford and of C. T. R. Wilson, had retained an interest in particle detection methods, although his main activity had been in gas discharges. He was to play a decisive role in the next big advance in particle physics (Figure 12.2). Here is his description of the events that led to the discovery of the pion:

In 1945 Powell was joined in Bristol by G. P. S. Occhialini who returned to England from Brazil before the end of the war. Occhialini was enthusiastic about the potentialities of the photographic method and made approaches to Ilford Ltd., with a view to improving the recording properties of the plates. Although it had already been shown that reliable range measurements could be secured, tracks could only be distinguishable if the number of developed grains in them was sufficiently numerous. This depends upon the number of ions knocked out of atoms by a particle per unit length of its path, a number which gets less the higher the velocity of the particle. In effect, as matters then stood, particles could only be detected when they were of relatively low speed, but most of the particles with speeds approaching that of light, which are most numerous in cosmic radiation, were not being recorded.

There were a number of ways in which it seemed possible that the recording properties of emulsions could be improved, by increasing the size and sensitivity of individual grains, for example, or by increasing the number of grains in unit volume of emulsion. C. Waller was the research chemist at Ilford's at that time and Ilford's methods of manufacture were such that he found it possible to make emulsions with a very substantial increase in the concentration of silver bromide. When the new emulsions were exposed and developed it was clear that a very remarkable improvement in performance had been achieved.

Late in 1946, Occhialini took a few small plates coated with the new emulsions—about 2 dozen each 2 cm × 1 cm in area, with emulsion about 50 microns thick—and exposed them at the French observatory in the Pyrenees at the Pic du Midi at an altitude of 3000 m. When they were recovered and developed in Bristol it was immediately apparent that a whole new world had been revealed. The track of a slow proton was so packed with developed grains that it appeared almost like a solid rod of silver, and the tiny volume of emulsion appeared under the microscope to be crowded with disintegrations produced by fast cosmic ray particles with much greater energies than any which could be generated artificially at the time. It was as if, suddenly, an entry had been gained into a walled orchard, where protected trees had flourished and all kinds of exotic fruits had ripened undisturbed in great profusion.

These new observations produced an atmosphere of the liveliest enthusiasm and anticipation in the laboratory. An intense search was begun of the small areas of the new emulsion which had been exposed and steps were taken to get more experimental material. By this time several microscopes for searching the plates were available with a number of girl observers, and a feverish hunt began. Almost everyday produced something new and exciting. At the beginning, the observers when they found any "event" in their search, such as a disintegration, would call a physicist to scrutinize it to see if it showed any remarkable features. Almost immediately Peter Fowler [Rutherford's grandson] who was in his final year as an undergraduate at that time, was shown an event in which, associated with a small disintegration, there appeared to be a particle, which from the characteristics of its track appeared to have a mass of about 200 m_e, which had reached the end of its range at a point where a disintegration occurred. There were only two possible explanations of the observed tracks. Either the particle had come to rest at a point which, by chance, coincided with that of a completely independent disintegration; or the tracks were related, in which case the sequence of events was unambiguous:—The particle of relatively small mass, a meson, must have reached the end of its range and produced a nuclear disintegration, when it was at, or almost at, "rest," with little or no energy of motion.

A few days earlier D. Perkins, at the Imperial College in London, who had independently been making similar experiments with the new emulsions, had found an "event" with closely similar characteristics. The observation of two "events" of a similar nature seemed to exclude completely the possibility of a chance juxta-position of unrelated tracks and it appeared certain that the consequences of the capture of a negative meson by a nucleus of an atom in the emulsion, and its resulting disintegration were being observed.

The observers soon learned to recognise the tracks of mesons and found many examples of similar disintegrations produced at the end of their range.

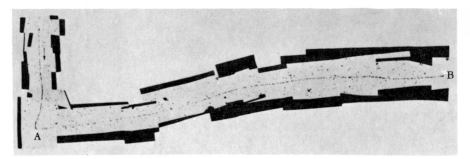

Figure 12.3 One of the first pictures of a pion. It was taken in a photographic emulsion by C. Lattes, G. Muirhead, G. Occhialini, and C. F. Powell in 1947. A pion stops at point A and here decays, emitting a muon. This demonstrates the relationship between pions and muons, both present in cosmic rays. (Courtesy of C. F. Powell.)

Indeed the lively interest of the observers was a crucial element in the progress of the work, and a good deal of trouble was taken to help them to learn to interpret the events they found and to understand the significance of what they were doing. [Seminario matematico e fisico di Milano, Simposio in Onore di G. Occhialini, Milan, 1959, p. 148]

The Powell-Occhialini group had thus found a meson that disrupted nuclei as required by the Yukawa hypothesis. Careful analysis of the tracks permitted them to conclude that the meson had a mass of about 139 MeV/c^2. Other tracks showed, however, that sometimes a meson decayed into a particle of about 106 MeV/c^2 and a neutral one with a mass very close to zero, presumably a neutrino.

The 106 MeV/c^2 particle decayed into one electron and more than one (presumably two) neutrinos. It was easy to identify the 106 MeV particle as the mesotron, while the new 139 MeV/c^2 particle that, on stopping, produces violent nuclear disintegrations instead of decaying freely, was identified as the meson postulated by Yukawa. Today, mesotrons are called *muons,* and the Yukawa-type particles are called mesons or *pions* (Figure 12.3). We shall use this nomenclature.

For the sake of clarity I show symbolically the reactions involved in the pion and muon decay, indicating the pion by π and the muon by μ: We consider a negative pion π; the case of the positive is similar.

$$\pi^- \rightarrow \mu^- + \bar{\nu}$$
$$ \longmapsto e + \nu + \bar{\nu}$$

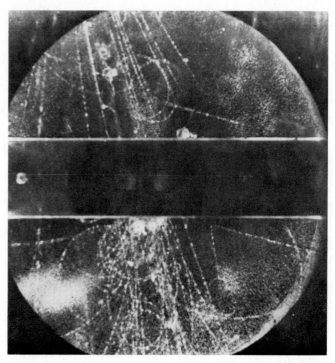

Figure 12.4 Cloud chamber picture taken by G. D. Rochester and C. C. Butler showing the first V particles from cosmic rays. For nearly two years these extremely convincing photographs were not followed by any others of the same kind. The V particle of this picture is now called K^0. [From *Nature 160*, 885 (1947).]

Both positive and negative pions may produce violent nuclear reactions, as pointed out previously. It is quite remarkable that all this could be read from a photographic plate with a microscope but without any other apparatus or, to be more precise, with only a tin can in which to expose the emulsions at high altitude. In time the technique was brought to a high degree of perfection by increasing the thickness of the emulsions, inventing appropriate methods of development, and loading the emulsions with other materials besides the photographically sensitive ones. The resulting refined technique made it possible to measure the velocity, mass, charge, and other characteristics of the particles that leave a track. For a while this technique dominated the study of elementary particles; it was later superseded by bubble chambers and other methods, although it is still useful and employed in special cases.

These discoveries stimulated an enormous amount of activity applying the same technique. The technique was very cheap, but it required hordes of scanners (almost all young women) willing to spend their days looking through microscopes searching for the tracks and measuring them. This activity was well suited for the European countries in the immediate postwar era, and in particular for Italy, which was suffering from great war damage and high unemployment. Thus Italy rapidly became an important

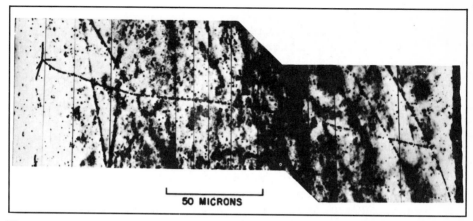

Figure 12.5 The first picture (in a photographic emulsion) of an artificial pion produced at Berkeley with the 184-inch cyclotron, taken by E. Gardner and C. M. G. Lattes. For the first time pions and muons were produced by an accelerator and not by cosmic rays. (Lawrence Berkeley Laboratory.)

center for this type of investigation, which is one of the reasons for the prominence given to particle physics in Italy.

A Horde of New Particles

Muons and pions are not the only new particles found in cosmic rays. In 1946, in Manchester, G. D. Rochester and C. C. Butler took many cloud chamber pictures of cosmic ray events and found in one of them tracks shaped like the letter V (Figure 12.4). The tracks could be explained only by admitting that they were generated by a particle of mass approximately 494 MeV/c^2 decaying in flight into two pions. The event remained unique for over a year, but the interpretation was so unequivocal that one had to believe in the existence of a new kind of particle, then called V, from the shape of the track. The particle of Figure 12.4 is now called a K^0 particle.

Photographic emulsions revealed other new particles in addition to pions and muons. For instance, the Bristol group found particles, called K^+ or K^-, that decayed into three pions, two positively charged and one negatively or vice versa. The list of new particles was lengthening when the new high-energy accelerators entered the field. They had reached energies sufficient to create first pions and later all the other particles discovered up to then in cosmic rays and many more.

The first machine with enough energy to produce new particles was the 184-inch Berkeley cyclotron (Figure 12.5). In 1948 E. Gardner, C. G.

Lattes (a Brazilian pupil of Occhialini who had participated in the discovery of the pion at Bristol and had moved from Bristol to Berkeley), and others used photographic emulsions as detectors and recognized the pions produced by the cyclotron. A little later B. Moyer and his students, again at Berkeley, detected gamma rays that were attributed to the decay of the neutral pion. This has a mean life of 0.8×10^{-16} sec in contrast with the mean life of the charged pions of 2.6×10^{-8} sec. The reason is that the neutral meson decays by electromagnetic interaction, a mode of decay forbidden to the charged mesons that must decay by the much weaker Fermi interaction. The neutral pion was the first particle discovered with the help of an accelerator. Most of the previous ones had been found in cosmic rays.

Cyclotron production of pions permitted the formation of pion beams and the performance of experiments that are totally inaccessible with cosmic ray sources. From that time on cosmic rays became instruments for the study of geophysics, cosmology, and other disciplines, but they lost importance for particle physics, where they could not compete with accelerators except at extremely high energies. Some time later other accelerators such as synchrotrons or betatrons succeeded in accelerating electrons to sufficient energy to produce mesons, either directly or by first producing photons.

Another leap in energy occurred when the Brookhaven National Laboratory activated the "Cosmotron," which, in 1952, surpassed 1,000 MeV, or 1 GeV. In 1953 this machine produced the V particles that had previously been seen in cosmic rays by Rochester and Butler; by this time, however, the situation had become complicated. Cosmic rays had been shown to contain, in addition to pions and muons, several types of particles heavier than nucleons. These particles decayed in various ways but always had neutrons or protons among their decay products. Such particles are called *hyperons*. Some of the most important with typical decays, are:

$$\Lambda \rightarrow p + \pi^- \ (1116) \qquad \Sigma^+ \rightarrow p + \pi^0 \ (1189) \qquad \Xi^- \rightarrow \Lambda + \pi^- \ (1321)$$

For the sake of clarity we use modern notation. The number in parenthesis is the mass in MeV/c^2. There are many variations of the decays written above. Furthermore, the particles may exist in different charge states.

In addition to hyperons cosmic rays were shown to contain the K particles already mentioned. They are heavier than pions but lighter than nucleons, and they decay in a variety of ways; for example:

$$K^+ \rightarrow \pi^+ + \pi^0 \ (494)$$
$$\rightarrow \mu^+ + \nu$$
$$\rightarrow \pi^+ + \pi^+ + \pi^-$$

The Λ particle presented a serious paradox. It was relatively easily formed, but it decayed very slowly. This fact contradicts some general principles of quantum mechanics that require that an easily formed particle also decays easily. There were several attempts to explain away this difficulty, but the true explanation was ultimately surmised independently by A. Pais and by K. Nishijima. The observed production and the decay process are not the inverse of each other as was believed, but entirely different, and thus there is no necessary connection between them. The Λ production occurs only in association with a kaon according to the equation

$$\pi^- + p \rightarrow \Lambda + K^0$$

while the Λ decay is given by the equation $\Lambda \rightarrow p + \pi^-$. The Λ production occurs always in association with that of a kaon, but its decay does not. The associated production depends on the strong Yukawa type interaction; the decay depends on the weak Fermi interaction.

In order to explain this "strange" phenomenon in 1953, M. Gell-Mann and T. Nakano and K. Nishijima independently postulated the existence of a new quantum number, called, for want of a better word, *strangeness*. Nobody knows if strangeness has some connection with other properties or with mechanical quantities such as the connection between angular momentum and certain quantum numbers, but it fixes things and explains several paradoxes besides the one mentioned above. We attribute a strangeness 0, ± 1, ± 2, and so on to particles and postulate that the sum of the strangeness of all particles involved does not change under the action of strong interactions. In this way we obtain selection rules that allow or forbid certain reactions, in agreement with experimental evidence.

Associated production and the strangeness phenomenology were clearly demonstrated in 1954, as soon as the Cosmotron reached sufficient energy for creating the associated particles. W. B. Fowler, R. P. Shutt, A. M. Thorndike, and W. L. Whittemore used to this end a diffusion cloud chamber that is continuously sensitive. They were soon followed by many other physicists employing a variety of techniques.

Murray Gell-Mann (Figure 12.6) is one of the aces of theoretical physics of the present generation. Born in New York in 1929, the son of a professor, he studied at Yale University and obtained his Ph.D. at age twenty-two at the Massachusetts Institute of Technology. Afterward he went to Chicago, where, like many other theoreticians of his generation, he came under the influence of Fermi. He is now professor at the California Institute of Technology, but he travels frequently and to many countries. For several of the important theoretical discoveries in the field of elementary particles Gell-Mann has taken the decisive step, simultaneously with but independently of other physicists: with Nakano and Nishijima for strangeness, with Y. Ne'eman for the eightfold way, and with G. Zweig for the idea of the quarks. Thus his existence may keep his theoretical colleagues and friends fearful of being scooped. At the same time, Gell-Mann occupies his excess brain power (a feature common to other theoreticians) with learning several languages, including Swahili, keeping abreast in biology, working on ecology, and advising the government on a variety of subjects.

In 1951 a synchrocyclotron capable of producing mesons started

Figure 12.6 Murray Gell-Man in 1961. He is one of the major theoreticians of particle physics, having introduced many of the new ideas into the field. (Photo E. Segrè.)

functioning at Chicago. Fermi had awaited it impatiently and had actively worked to hasten its completion. Now, he could finally return to experimental work for which he had so carefully prepared. His efforts were compensated; shortly before his death he found a new and unexpected phenomenon. The collision cross-section of positive pions with protons showed a huge maximum. This was a clear sign of the formation of semistable compound particles, sometimes called *resonances,* or simply *particles.* After Fermi's initial find hundreds more were recognized, and their study is still an active field of investigation.

Antinucleons

A little later the possibility arose of giving a decisive answer to a long-standing question. As the reader may recall, Dirac had predicted the positron, which was later discovered by Anderson. By simply extending Dirac's theory to protons, one would expect the existence of an antiproton, of equal and opposite charge and of the same mass as the proton. However, such simple extrapolation may be unwarranted, as was clearly shown, for instance, by Stern's discovery that the magnetic moment of the proton is very different from what one would naively guess on the basis of Dirac's theory. Thus it was important to give a clear yes or no answer on the existence of the

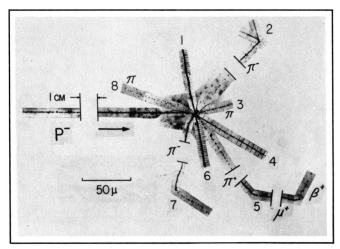

Figure 12.7 Antiproton annihilation in a nuclear emulsion. An antiproton (p^-) stops and is annihilated with a nucleon from another nucleus. Pions and other particles move away from the point of the collision. The energy of the visible particles is greater than the energy at rest of the antiproton, which demonstrates that there exists another particle that is annihilated. (Lawrence Berkeley Laboratory.)

antiproton. Observation of cosmic ray events had given some indications on the subject, but no clear answer.

In 1955 the Bevatron at Berkeley reached the energy of 6 GeV, equal to about 2 GeV in the center of mass. This was the minimum required to produce a proton-antiproton pair, if the antiproton existed at all. O. Chamberlain, C. Wiegand, T. Ypsilantis, and I succeeded in demonstrating its existence convincingly. One could thus be sure of the possibility of antimatter, indeed even of whole antiworlds, although so far we do not know whether such antiworlds really exist.

The symmetry between particle and antiparticle is one of the new truths of physics. For each particle there is an antiparticle of equal mass and opposite electric charge, of equal and opposite strangeness, of equal spin, of equal and opposite magnetic moment. In short, all properties are either equal or opposite. If we replace every neutron and proton in a nucleus with an antineutron and an antiproton, we have an antinucleus. We can now dress it with antielectrons—that is, positrons—and we have an antiatom. With antiatoms we may form antimolecules, and so on up to entire antiworlds. For an antiman living in an antiworld everything would look and behave in the same way that it does for us. There are no intrinsic differences between a world and an antiworld, and astronomical observations cannot tell us if a star is of matter or antimatter. However, if matter and antimatter meet (Figure 12.7), they annihilate each other, and in a very short time all their energy is transformed into neutrinos, antineutrinos, and gamma rays that flee the place of the annihilation with the speed of light.

The Downfall of Parity

In contrast to the more or less expected discovery of the antiproton, at about the same time a supposedly safe tenet that had dominated much of physics was disproved. If we perform any physics experiment and look at it either directly or by reflection in a perfect mirror, there is no way of telling whether we are looking at the experiment directly or through the mirror. It is true that if we look at a man directly or in a mirror, we can tell whether his jacket is buttoned correctly or backward. Similarly, we can tell whether a screw is right-handed or left-handed, but these are arbitrary conventions and not laws of nature. Even a man with his heart on the right side could function perfectly well. The reader may think that there are some laws of electricity that contain an intrinsic right- or left-handedness, but careful analysis will show that this is not so; the handedness is connected with a convention on the sign of the electric charge. All strong and electromagnetic interactions obey this reflection symmetry exactly. The fact that a phenomenon and its mirror image are either both possible or both impossible is called the *conservation of parity*.

Whenever a regularity is valid in many cases, there is a tendency to generalize it to other untested circumstances and perhaps even to make it into a "principle." If possible, one frosts the cake with some philosophical considerations, as happened with the concepts of space and time before Einstein. Such occurred with the principle of conservation of parity, and when it failed experimentally, physicists such as Pauli were deeply shaken.

The events that led to the discovery of the nonconservation of parity started in 1955, with the observation of the decay of certain K particles, then called θ and τ. The two particles had the same mass according to experimental measurements, and they had the same mean life, but they decayed in different ways. The natural assumption was that they were the same particle decaying in different modes. There was nothing unusual in this. It was known since the time of Curie and Rutherford that just as people can die of different diseases, radioactive atoms may decay in alternate fashions—for instance, emitting beta or alpha particles. But the strange fact about the K decay was that it did not seem to obey the conservation of parity. Conservation of parity forbids one particle from decaying by the emission of two pions or, alternately, three pions, and this was exactly what seemed to occur in the case of the θ and τ. This serious dilemma was resolved in a completely unexpected way by the Chinese physicists Tsung-Dao Lee and Chen Ning Yang. They pointed out that there was no direct proof that weak interactions, those discovered by Fermi in 1933 and responsible for beta

Figure 12.8 Chien Shiung Wu (right) and W. Pauli. Wu, who did fundamental studies on weak interactions, is photographed here with the inventor of the neutrino hypothesis. The neutrino plays an important role in weak interactions.

decay, conserve parity. If a weak interaction does not conserve parity, the dilemma vanishes. They also indicated experiments that could verify the hypothetical nonconservation of parity. The situation somewhat resembled the story of the Emperor's new clothes. Two swindlers obtained golden threads to weave a coat for the Emperor, saying it would be visible only to intelligent people. Everybody was ashamed to admit that he could not see the coat and all praised its beauty. Only when a child at the parade exclaimed: "The Emperor is naked" did the truth emerge.

Immediately after the publication of Lee and Yang's paper a group of physicists from Columbia University and from the National Bureau of Standards in Washington, D.C., led by another Chinese physicist, Chien-Shiung Wu (Figure 12.8), and two other groups independently demonstrated that the conservation of parity was a myth in the case of weak interactions.

Wu's group found that nuclei of radioactive cobalt oriented so as to have their spins all pointing in one direction tended to emit electrons parallel rather than antiparallel to that direction. If parity were conserved, the parallel and antiparallel directions should be equally probable. For the emission of gamma rays, an electromagnetic phenomenon, this is the case, but the beta emission, due to weak interactions, does not conserve parity.

Who are these Chinese physicists? Chien-Shiung Wu was my first student at Berkeley, where she had arrived from her native Shanghai. Her will power and devotion to work are reminiscent of Marie Curie, but she is more worldly, elegant, and witty. She has devoted most of her scientific work to the study of beta decay, in which she has made several important discoveries. She is married to another Chinese physicist, Luke Yuan, and is the mother of a young physicist, Vincent.

Chen Ning Yang was born in Hofei, Anwei, in 1922, the first of five children. His father was a professor of mathematics at Tsinghua University near Peking. Both Yang and his friend Tsung-Dao Lee studied at National Southwest University in Kunming. After the war Yang came to the United States hoping to do graduate work under Fermi, but he did not know where Fermi was. He went to Columbia University, only to find out that Fermi had moved to Chicago; he then joined him there. He first did experimental work without much success, but after a while his theoretical ability showed up. In 1946 he wrote his Ph.D. dissertation under Teller, but later he worked with Fermi.

Lee was born in Shanghai in 1926 and attended school there. At the university he met Yang. His professor, Ta-You Wu, secured a fellowship for him to study at Chicago, where he obtained his Ph.D. in 1950 with Fermi. After obtaining their doctorates, both Lee and Yang worked in several major American universities and collaborated for many years. In addition to the historical paper on the nonconservation of parity, they obtained important theoretical results in a variety of subjects ranging from statistical mechanics to field theory (Figure 12.9). At present Yang is at the State University of New York at Stonybrook, and T. D. Lee is at Columbia University.

This trio of Chinese physicists shows what China's future contribution to physics could be if that great country overcomes the period of revolutionary convulsions and resumes its historic role as one of the leaders of civilization, as witnessed by the early European travelers, to their astonishment.

Two other groups recognized the nonconservation of parity in the muon decay—A. M. Friedman and V. Telegdi at Chicago, using photographic emulsions; and R. L. Garwin, L. M. Lederman, and M. Weinrich at Columbia, using electronic means. They found a correlation between the direction of the muon spin and the direction of the emission of the decay electrons. Immediately after the original discoveries mentioned above, many scientists rushed to investigate all the alternatives opened up by the nonconservation of parity, and in a short time hundreds of papers on the subject were published. With almost equal rapidity, experimental results decided between the alternatives proposed and thus brought about a satisfactory clarification of beta decay.

Figure 12.9 Eight Nobel Prize winners at the 1960 Rochester Meeting at Rochester, N.Y. From the left, E. Segrè, C. N. Yang, O. Chamberlain, T. D. Lee, E. McMillan, C. D. Anderson, I. I. Rabi, W. Heisenberg. These meetings were important periodic events for particle physicists. They gave one of the first opportunities for scientific exchanges between Western and Soviet physicists. (Photo L. Cuzer.)

As stated above, parity violation permits us to distinguish a phenomenon from its mirror image (Figure 12.10). However, there is a further surprise that reestablishes the symmetry, albeit at a different level. The mirror image of a phenomenon is exactly what would be seen if we performed the experiment with antimatter. From a real situation we pass, by reflection in a mirror, to a situation that we call P transform. Replacing every particle with its antiparticle, we obtain a C transform. By performing both transformations, we obtain a PC transform; a CP transformation, in which we first change a particle into its antiparticle and then reflect it in a mirror,

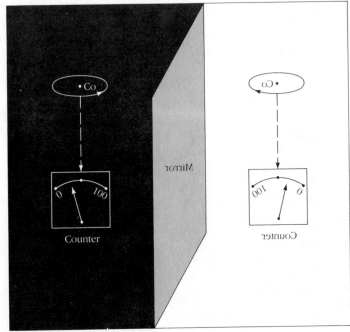

Figure 12.10
Decay of Co^{60} and its mirror image. [From C. N. Yang, *Elementary Particles, A Short History of Some Discoveries in Atomic Physics* (copyright © 1962 by Princeton University Press), Fig. 36, p. 60. Reprinted by permission of Princeton University Press.]

gives the same result. If we start from a real situation, a CP or PC transformation brings us again to a real situation.

In addition to C and P transformations, we can also consider T transformations, in which all velocities are inverted. The resulting phenomenon is as if we were looking at a movie film of the experiment run backward. Performing all three transformations—C, P, and T,—from a real phenomenon we again obtain a real or possible phenomenon. This fact is very dear and important to theoretical physicists who can prove it on very general assumptions such as relativity. The discoveries on nonconservation of parity showed that CP is a valid transformation by itself, and thus if CPT is valid, then T must also be valid by itself.

It would be convenient if the story ended here, but to be exact there is still a further complication. In 1964 J. H. Christenson, R. Turlay, V. L. Fitch, and J. W. Cronin found in the decay of neutral K-mesons a case in which the CP symmetry is violated. It is a small but indisputable effect. On the other hand, it seems that CPT invariance is preserved by a corresponding violation of T that compensates for that of CP. Here we are at the limits of what is known.

The violation of parity conservation is perhaps the greatest theoretical discovery of the postwar era. It has extinguished a prejudice that had been transformed into a principle on the basis of insufficient tests.

As a footnote, let me add that an American physicist, R. T. Cox, may have observed the nonconservation of parity in beta decay as early as 1928. Using decay electrons in double scattering experiments, he had observed certain asymmetries that pointed to a polarization of the initial electrons. This would have been in conflict with the conservation of parity, but he was so convinced that parity was conserved that he repeated the experiment with thermionic electrons, which are not polarized. With the thermionic electrons the effect vanished because the polarization does not occur, and he interpreted the result by assuming that the first part of the experiment was somehow in error. Here is one more case of a discovery missed because the investigator was not mentally prepared for a surprising result.

The Bubble Chamber

We have seen that Fermi found the first resonance in the proton-pion system, but nobody suspected then that there would be so many resonances and in all systems. The invention of the bubble chamber technique opened the floodgates. First a few resonances were found in the proton-antiproton annihilation (Figure 12.11), then dozens, then hundreds appeared in all kind of reactions. The hunt for resonances kept physicists busy in the 1960s, and it is still going strong.

For some time the new technique, the bubble chamber, has dominated the study of resonances, just as earlier the photographic emulsions technique had dominated particle physics. The bubble chamber is an instrument that shows the paths of charged particles. We have already seen that, as early as 1910, C. T. R. Wilson had demonstrated the tracks of charged particles moving in saturated vapor by revealing the condensation along them. A sudden expansion of the chamber produces supersaturation in the vapor, and the liquid condenses preferentially on the ions left on their path by charged particles. A strong side illumination makes the fog visible, just as we can see the vapor wake left by airplanes flying at high altitudes. The Wilson cloud chamber has a glorious history; it has revealed, among other things, the tracks of the first artificial disintegrations; the recoil protons set in motion by the neutrons; the positron; and the showers. However, it has the great drawback that the gas has low density; in other words, it contains very little matter per unit volume.

In 1952 Donald A. Glaser (b.1926) thought of replacing the gas with a liquid, thus increasing the density by roughly a factor of a thousand. He used a liquid at its boiling point and observed the bubbles formed on the path of the ions when the pressure was suddenly lowered and the liquid thus brought above its boiling point. Glaser developed the art starting with small chambers full of ethyl ether. Once he had succeeded in observing the first tracks, he tried different liquids. The most important ones were liquid hy-

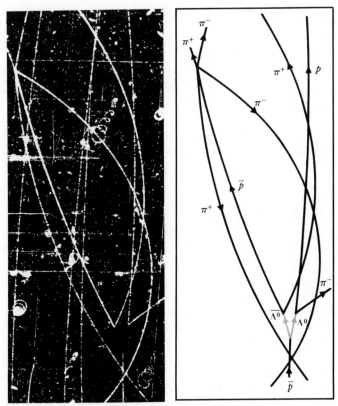

Figure 12.11 Picture taken in a hydrogen-filled bubble chamber of the production and decay of a pair of neutral lambda-antilambda particles. The Λ on the right in the schematic drawing decays into a proton (p) and a negative pion (π^-). The antilambda (on the left) decays into an antiproton ($\bar{p}$) and a positive pion (π^+). On hitting a proton in the liquid hydrogen, the antiproton is annihilated into four pions. (Lawrence Berkeley Laboratory.)

drogen, which gives a simple target, and xenon, which gives a target with high atomic number. Glaser led the way in this field but later turned to biology. His bubble chambers were increased in size and scope by another physicist.

L. W. Alvarez (b. 1911) is one of the physicists who grew up under the influence of Lawrence, thus acquiring a tendency toward big experimental enterprises. A man of great imagination, he obtained several important results in cosmic rays with his teacher A. H. Compton at Chicago, but he soon moved to Berkeley and worked on the cyclotron. During the war Alvarez went to MIT at Lawrence's prompting and there invented a system for landing aircraft under radar guidance in conditions of poor visibility. From MIT he moved to Los Alamos and later insisted on flying in the airplane that dropped the bomb on Hiroshima. Alvarez had a strong urge to be present at events that he thought would be historic. He liked to meet important people and felt a sort of hero worship for great scientists. On his return to Berkeley after the war, he planned and directed the construction of a linear accelerator and then became involved in the Livermore enterprise.

When Glaser invented the bubble chamber, Alvarez, convinced of its great possibilities, decided to build one of unprecedented dimensions filled with liquid hydrogen. This was a major technical enterprise—even Lawrence had some qualms about it—but Alvarez enlisted a large group of technical experts in various fields, aroused their enthusiasm in the Lawrence tradition, and proceeded to build a series of bubble chambers of increasing size until he reached one 72 inches long, the size of a large bathtub. Now the biggest bubble chambers are several meters in diameter and contain tens of thousands of liters of liquid hydrogen. Alvarez's contained "only" 500 liters.

Such instruments are very complex and very expensive; their cost is comparable to that of accelerators. Nevertheless, they are essential to the exploitation of accelerators, and they are among the most effective tools for the study of elementary particles. For full advantage of their power, millions of pictures must be scanned rapidly by semiautomatic methods. The output of the scanning devices is fed into computers and analyzed. The programming of the computers for this job was one of the difficult aspects of setting up the system. To appreciate the progress, one should compare Blackett's direct analysis of the stereoscopic views of his cloud chamber pictures containing the nitrogen disintegration (p. 111) with the modern operation of a bubble chamber. The wedding of the bubble chamber to the computer has been prolific, and film obtained in the big laboratories is now distributed all over the world to users who study it again and again, mining results out of the raw material.

Order in the Wilderness

Bubble chambers, photographic emulsions, and other more modern techniques such as spark chambers supplement each other and are combined in ingenious ways to accumulate experimental information on elementary particles.

The data obtained recall the atomic energy levels, and their classification and systematization is an obvious task for particle physics. Although the task is obvious, the means for accomplishing it are not clear. Fermi and Yang in 1949 had tried a composite model based on two fundamental constituents: proton and neutron (with their antiparticles). With the discovery of strangeness three objects became necessary. S. Sakata and his pupils (M. Ikeda, S. Ogawa, and Y. Ohnuki) starting from 1955 developed a model based on neutron, proton, Λ, and their antiparticles, including its peculiar mathematical treatment. However, their work did not correspond to the facts. With the accumulation of the empirical results, at a certain point a solution occurred independently to both Gell-Mann and Ne'eman. Gell-Mann circulated a preliminary version of his paper on January 20, 1961. Ne'eman, unaware of Gell-Mann, had reached approximately the same conclusions and submitted a paper for publication on February 13, 1961.

Yuval Ne'eman (b. 1925), an Israeli physicist, had been an active officer in the Israeli army and in a period of peace had been assigned as military attaché at the London embassy. There, in his spare time, he decided to study physics and mathematics. He became a pupil of Abdus Salam, a Pakistani physicist who taught at the Imperial College in London. Ne'eman, who is a strong mathematician, soon recognized that there was a relation between the groups of elementary particles known experimentally and a mathematical theory called group theory. In particular, the hadrons are connected to a special group called (SU 3). I mention the name without entering into technical details; suffice it to say that with the help of this mathematical scheme it is possible to classify the particles into families in a way reminiscent of Mendeleev's periodic system. The observed regularities are so clear as to allow the prediction of missing particles, predictions fulfilled by subsequent experiments. In its time the same happened with Mendeleev's periodic system; the chemist had no theoretical basis for his table, only a well-deserved confidence in his extrapolations. The explanation of Mendeleev's system required a thorough knowledge of atomic physics that had taken about sixty years to develop. To a certain extent the same is true for the (SU 3) classification. We may, however, be much closer to a substantive explanation of (SU 3), through the quarks.

The abstract mathematical results of (SU 3) may be obtained by postulating the existence of subunits called *quarks* by Gell-Mann, who invented them at the same time as George Zweig. The name *quark* shows Gell-Mann's familiarity with James Joyce's *Finnegan's Wake,* where the term occurs. Free quarks have never been seen. They would have properties that would make them easily recognizable—for instance, the electric charge one-third or two-thirds times the charge of the electron or proton. Physicists have looked carefully for quarks everywhere from moon rocks to cosmic ray showers, but to no avail. Nevertheless, the quark hypothesis accounts for many things besides (SU 3)—for instance, masses, magnetic moments, cross-sections, and so on—so that it carries considerable weight. Even if the quark hypothesis is only a temporary stage in the development of particle theory, we may recall that temporary steps, such as the Bohr modellistic theory, can generate great progress.

Theoreticians have found reasons why free quarks could be prevented from manifesting themselves. In classical physics free magnetic poles do not exist, and if we cut a magnet we obtain two complete magnets, not two separate poles. This might be analogous to the quark situation. For instance, a pion is supposed to contain a quark and an antiquark. If we try to separate them, the energy employed generates a new quark-antiquark pair that at the moment of separation attach themselves to the original quark and antiquark respectively. They produce thus two pions but no free quark: Apart from this difficulty, the initial quark hypothesis contemplated three of them with their antiquarks. The three quarks were called *u, d,* and *s,* for *upward, downward,* and *sideways.* They were intimately connected to the three conserved quantities: baryon number, electric charge, and strangeness. Conserved means that no reaction can change these three quantities; they must be the same on both sides of any equation representing a natural phenomenon. Weak interactions, however, can change strangeness but not the other two quantities.

The quark hypothesis is most attractive, but in recent years it has started to show disturbing symptoms. First, it was found that the quarks had to have an additional quantum number, called *color,* that can take three values. Color is needed to prevent a horrible breach in the structure of quantum mechanics, because without it the relation between spin and statistics (see p. 155) and other important tenets would fail. Thus the quarks would be nine, and of course each one should have an antiquark. This, however, is not the only problem.

In 1972 Sheldon Glashow, T. Iliopoulos, and Luciano Maiani, three young theoreticians, noted that some phenomena that did *not* occur would be forbidden by the existence of a fourth quark having some special properties, which they specified in detail. The argument for a fourth quark was very indirect, although it was fairly precise, if at all valid. Two years later, a whole family of new and most remarkable particles, discovered in a period of a few weeks, could be explained by the fourth quark, which thus acquired solid support.

Samuel C. C. Ting, who led one wing of the experimental advance, was born, prematurely, in 1936 in Ann Arbor (Michigan) while his Chinese parents were visiting the United States. When he was two months old, his family went back with the new child to China, where he remained until the age of twenty, at which time he returned to the United States to complete his education at the University of Michigan. Ting was especially interested in what he calls *heavy photons,* particles that would decay in an electron-positron pair, and he looked for them among the products of high-energy collisions of protons or beryllium. Such collisions produce all kind of particles, and the electron-positron pairs are rare indeed—1 in every 10^6 collisions. Their signature, as physicists call the visible tracks by which a particle may be recognized, is very characteristic and difficult to forge. Ting and his colleagues saw the looked-for particle for many months but kept the result secret in an effort to close all loopholes by which it could be simulated. Finally they were ready to publish their result, but a surprise awaited them.

At Stanford, California, a group of SLAC (Stanford linear accelerator) physicists in cooperation with a group of Berkeley physicists had been observing the same particle for some time. However, they formed it by a different reaction: electron-positron collisions. These had become possible by the development of an ingenious scheme in which electrons and positrons are accelerated in opposite directions and hurled one against the other (the center of mass of the two particles remains at rest). Such colliding beam accelerators had been fed by the large linear accelerator of SLAC. Burton Richter, one of the moving spirits behind this enterprise, was also one of its leading users, once it was built. The SLAC physicists noted that the electron-positron collisions generated more hadrons than expected. In due course an energy for the colliding particles was singled out, for which this phenomenon took completely unexpected proportions, and ultimately a sharp resonance was located that corresponded to a particle of a most unusual long life for its energy. This occurred within days of Ting's final certainty about *his* particle, and it was apparent that they were the same. The December 2, 1974, issue of *Physical Review Letters* carried two letters from the discoverers. Unfortunately, they could not agree on the name for the new particle. Ting called it J; the Stanford physicists, ψ. We will follow the latter, to avoid a double name. The same issue of *Physical Review Letters* also contained a third letter of confirmation and accurate measurements on the ψ that were obtained at Frascati, Italy with another colliding electron-positron beam apparatus. The Roman physicists had been informed of the discovery by telephone; their machine had a nominal energy below the threshold for ψ formation, but they pushed the machine to a higher energy, and by straining it to its limits, they were able to exceed the threshold for ψ formation.

The peculiarity of the ψ is that it has a large mass—3,098 MeV—and a life of 10^{-23} sec. This is a short time, but it is at least 1,000 times longer than what could reasonably be expected on the basis of the available energy, if the particle decayed by strong interaction. The puzzle of the relative stability of the ψ is similar to the puzzle of the stability of strange particles. Indeed, the reason is similar and the relative stability of the ψ is attributed to a new quantum number called *charm,* which is conserved in strong interactions. But a new quantum number requires a new quark; lo and behold, this was exactly the quark hypothesized by Glashow and his friends. The original ψ was soon followed by a whole family of particles containing charmed quarks. Thus quarks are of at least four species. I say at least because experiments by Lederman and others in 1977 already seem to require an additional quark. Maybe as we increase the energy, more quantum numbers appear. This would not be a very cheerful prospect if we ever hope to achieve a closed theory.

Quarks are the basic components of the strong interacting particles hadrons, but leptons, the particles subject to electromagnetic and weak interactions only, have also shown a tendency to proliferate. At first there were only two types of leptons: electrons and neutrinos. In the late 1930s

Figure 12.12 Steven Weinberg (left) and Abdus Salam (right) striving to explain their theories to audiences of physicists in 1979. (Photos by E. Segrè.)

muons were added. Subsequently it was shown that neutrinos are of two different kinds: one connected to electrons and the other to muons. They differ because they produce different reactions: One generates electrons, the other, muons. But this is not all. There are experiments that indicate the existence of heavier charged leptons, and they even may have other neutrinos as companions. Martin L. Perl and collaborators who have seen this lepton at SLAC, call it τ, for $\tau\rho\iota\tau o\nu$, the third lepton.

From the theoretical point of view there are serious attempts to develop comprehensive theories that would unify weak and electromagnetic interactions and possibly even strong ones. These theories would create a higher synthesis just as electromagnetism unifies electricity and magnetism. Steven Weinberg and Abdus Salam independently formulated a possible theory. This theory is not unique and is not proved, but it has shown considerable predictive power and coordinates several of the latest discoveries in a remarkable fashion. Weinberg is a red-haired professor of theoretical physics at Harvard, who exercises his excess brain on medieval history. We already met Salam, the Pakistani professor at London. He now also directs a center for theoretical physics at Trieste (Italy), mainly for the benefit of scientists of underdeveloped countries (Figure 12.12).

The whole field of particle physics is thus in motion and could suddenly explode in some great discovery. For instance, the race is on for finding some very heavy particles (80 GeV/c^2) predicted by the Salam–Weinberg theory.

I shall not expand further on particle physics, but rather turn to some other aspects of contemporary physics, although they are far from my work.

Chapter 13

New Branches
from the Old Stump

In this chapter I would like to give an idea of what has happened during the postwar era in branches of physics that I have not previously mentioned. First, the amount of research has increased so much that further specialization has become necessary. Consequently, there are a number of new journals devoted to ever-narrower fields. Second, in the postwar physics I do not see any towering figures like Rutherford, Einstein, or Bohr.

The *Physical Review,* published in the United States, is the largest periodical that attempts to cover the whole spectrum of physics; it is subdivided into five sections: general physics and theory, atomic physics, nuclear physics, solid-state physics, and particle physics. There are practical reasons for this division; individuals seldom subscribe to more than one or two sections, and the annual volume of the journal, including all five sections (about 30,000 pages), could hardly be accommodated in modern houses or offices. Statistics on the professional occupations of physicists show that research in these five fields absorbs most physicists, excluding those who only teach or those who work on applications verging on engineering. The classification, like all such classifications, is somewhat conventional, and there are extensive overlaps and undefined boundaries. Since it is not possible to give a complete description of all these fields, I will mention highlights chosen primarily on the basis of my personal knowledge or preference.

Quantum Electrodynamics

It often happens that when conditions are improved by some new experimental or theoretical technique, old subjects give new and unexpected results (see, for instance, Zeeman's discovery, p. 13.) New important facts coming to light show how incomplete previous knowledge was, or technical advances sometimes permit feats that previously would have been considered science fiction.

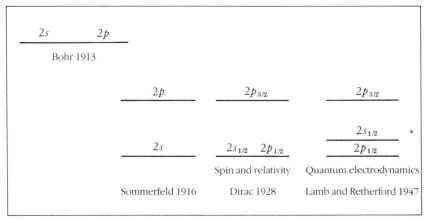

Figure 13.1 The historical development of the theory of energy levels in a hydrogen atom. The example used is the second level, whose total quantum number is equal to 2. According to Bohr (1913), it consisted of two orbits, a circular one (level $2s$) and an elliptical one (level $2p$), both with exactly the same energy. According to Sommerfeld (1916), relativity changes the energy slightly and separates the $2p$ level from the $2s$ one. G. E. Uhlenbeck and S. A. Goudsmit (1925) introduced the spin and the electron magnetic moment. The two levels are now 3, that is, $2s_{1/2}, 2p_{1/2}, 2p_{3/2}$, but if one includes relativity, the first two coincide, leaving only two distinct levels, as in Sommerfeld's previous theory. Dirac's theory (1928) does not vary the position of the levels and their quantum numbers, but it links spin with relativity. In 1947 W. E. Lamb and R. C. Retherford found a separation between $2p_{1/2}$ and $2s_{1/2}$ (called the Lamb shift), which is explained by quantum electrodynamics.

A beautiful example comes from the most classical spectroscopy. The fine structure of the hydrogen levels had been studied many times by optical means. Some spectroscopists had even hinted that all was not well and that they did not agree perfectly with Dirac's theory, but these voices were feeble and the general opinion was that the agreement was perfect. Experiments contrary to this view could never be corroborated.

After the war Willis E. Lamb (b.1913) and his student, R. C. Retherford, using much more refined methods than those of optical spectroscopy, demonstrated by a combination of molecular ray and microwave techniques a serious discrepancy in Dirac's theory: The $s_{1/2}$ and $p_{1/2}$ levels that should have coincided were separated in energy (Figure 13.1).

Almost at the same time Polykarp Kusch (b.1911) showed that the magnetic moment of the electron, which should have been exactly one Bohr magneton, was slightly larger. These discoveries were announced at a Shelter Island Conference organized by Oppenheimer in 1947. This conference and a few held later were a sort of substitute for the Solvay Councils, which had been disrupted by the war. The general theme of the conference of

1947 was quantum field theory. The application of quantum principles to the electromagnetic field was the prime example. There had been many attempts in this direction by the founders of quantum mechanics, but success had been only partial. Infinite quantities appeared in the theory and prevented the attainment of rigorous results, although it was known more or less how to eliminate the disturbing infinities practically, if not in a mathematically acceptable fashion.

Substantial progress had been made during the war in Japan by Sinitiro Tomonaga (b.1906) and his associates, but this was not yet known in the West. When the experiments of Lamb and Kusch were reported, the participants at Shelter Island entered into a protracted and intense discussion of the underlying theory. On the train returning to Cornell University, Hans Bethe made a first calculation of the Lamb shift that, although imperfect, gave a clue to the phenomenon. Among the participants in the Shelter Island Conference were the two young theoreticians J. Schwinger (b.1918), a former Rabi protégé; and R. P. Feynman (b.1918), who had worked at Los Alamos (Figure 13.2). In the following months they reformulated quantum field theory in a form very suitable for calculation, and with better definitions of "renormalized" masses—the real parameters accessible to observation after the infinite masses of the theory have been reduced by subtraction of another infinity.

Figure 13.2 From the left,
R. P. Feynman, J. Schwinger, and
S. Tomonaga, who together won the
1965 Nobel Prize for their research in
quantum electrodynamics. (Photos left
and center E. Segrè; right American
Institute of Physics, Meggers Gallery
of Nobel Laureates.)

The present agreement between theory and experiment is admirable, perhaps the most precise in all physics. For instance, the magnetic moment of the electron can be measured to 2 parts in 10^9, and the result of the measurement agrees with the calculation to 1 part in 10^9. Also, the calculation has a residual uncertainty of about 3 parts in 10^9 because of uncertainties in higher approximations. In spite of this remarkable achievement the theory still has logical flaws deriving from the subtraction of infinities. This is the opinion of Dirac and of most theorists.

In his formulation of quantum electrodynamics Feynman invented a powerful calculation technique now familiar to most theoretical students. He used the so-called Feynman diagrams that represent the phenomenon in an intuitive way and at the same time permit analysis and derivation of mathematical formulae. Feynman's method resembles a shorthand writing that summarizes very long and laborious calculations, in which it would be easy to incur algebraic errors, and transforms them into much simpler procedures, obeying a fixed set of rules. When Feynman's papers appeared, many learned his techniques and used them fluently without really understanding their justification. My friend G. C. Wick (b.1909), a distinguished theoretical physicist and then professor at Berkeley, was incensed by this state of affairs; he found that it was too much to have professors and students of theoretical physics who could give him all the rules without any true

Figure 13.3 The Solvay Council of 1961. Seated, from left to right: S. Tomonaga, W. Heitler, Y. Nambu, N. Bohr, F. Perrin, J. R. Oppenheimer, Sir W. L. Bragg, C. Møller, C. J. Gorter, H. Yukawa, R. E. Peierls, H. A. Bethe. Second row: I. Prigogine, A. Pais, A. Salam, W. Heisenberg, F. J. Dyson, R. P. Feynman, L. Rosenfeld, P. A. M. Dirac, L. van Hove, O. Klein. Third row: A. S. Whightman, S. Mandelstam, G. Chew, M. L. Goldberger, G. C. Wick, M. Gell-Mann, G. Källen, E. Wigner, G. Wentzel, J. Schwinger, M. Cini. There is a mix of the old generation: Bohr, Bragg, Dirac, with the newer forces especially active in quantum electrodynamics: Tomonaga, Schwinger, Heitler, Bethe, and some whose main work lay in the future: Prigogine (thermodynamics), Salam. (Institut Solvay.)

proof. He then proceeded to establish a sound basis for the mathematical theory and wrote some papers on the subject that have become classics. In 1949 F. Dyson showed that the methods of Schwinger and of Tomonaga were bound to give the same results as Feynman's. Most of these physicists can be seen in Figure 13.3.

Fermi had developed the theory of weak interactions in 1934, modeling it on quantum electrodynamics, as it was known at that time. It turns out that Fermi's theory is not renormalizable, although it gives excellent results in first approximation because of the weakness of the interaction. When one tries to make a field theory of strong interactions, following Yukawa, the

difficulties become insurmountable because the method of successive approximations fails badly, given the magnitude of the strong interaction. These difficulties discredited field theory in the 1960s, but it has now been revived and is again in favor, expecially because it seems to offer a possibility of unifying the theories of the different interactions. There will be experimental tests that may validate or shatter these hopes.

To an outsider this ferment resembles the pre-Einsteinian studies in electricity in moving systems. There may be fundamental defects that will require radical cures. Once in a while one sees theories that become fashionable for a few years and then fade away, leaving some partial results of permanent value. In any case, this is not a new situation in physics.

Laser and Maser

In the 1920s we used to joke that good physicists, once passed to their heavenly rewards, would find apparatus in paradise which, with a twist of some knobs, would give electromagnetic radiation of any desired frequency, intensity, polarization, and direction of propagation. In 1917, when Einstein wrote the paper quoted on p. 95, he did not anticipate that he was preparing the conceptual means of achieving the reward for good physicists, but this is in fact what happened. Lasers, masers, and similar apparatus come very near to the practical realization of the ideal apparatus, and the idea of stimulated emission, described in Einstein's scheme by his B coefficient, is basic to all of them.

These instruments were built after the war, almost always by veterans of radar work who used many of the arts they learned and developed during the war for scientific purposes later. In 1954 Charles H. Townes (b.1915) was the first to build an amplifier that used excited molecules as the active element, in this case, ammonia. The art was developed by many workers all over the world and has led to lasers and masers that generate beams of unprecedented intensity and monochromaticity that are also collimated to an enormous degree. As an example of the feats that can be achieved with such beams, it has been possible to place mirrors on the moon and detect laser beams sent from the earth and reflected on them. The laser has brought about a revolution in optics and has permitted the observation of new phenomena, such as those of nonlinear optics, and the development of new arts, such as holography.

Nuclear Physics

When we were working in nuclear physics in Rome in 1934, I prepared a chart having the atomic number as abscissa and the number of neutrons contained in a nucleus as ordinate. I marked all known nuclei with dots: black for stable nuclei, red for radioactive ones. Every week we were proud to be able to add some more red dots. A few years later, I found a similar chart in Berkeley, in which the dots had been replaced by key tags hanging from nails. The chart occupied a whole wall, and the key tags were changed as new information was uncovered, which happened frequently. At the end of the war, in Los Alamos, in a period of respite from previously hectic work, I decided to update a similar chart. By then, however, the numbers to be reported were in the thousands, and I had to enlist the help of my wife, Elfriede, for the task. In addition to the old data—mass, mean life, decay modes—there were innumerable new data, such as energy of the emitted radiations, neutron capture cross-sections, spins, and so on. The resultant table had great success and was printed in tens of thousands of copies. The security people at Los Alamos raised their brows, but in the end they permitted publication. Now similar tables are produced by elaborate organizations with many employees, sometimes connected with nuclear industries, and are often used as backdrops in portraits of admirals commanding nuclear vessels or fleets. A recent compilation (1979) of nuclear data fills over 1500 pages of a book that resembles a telephone directory.

The study of such tables is a fascinating pastime and a valid source of ideas for the expert. The accumulation of observational material has permitted the discovery of previously unknown regularities and thus the development of nuclear models of considerable detail and remarkable predictive power.

One of the most successful models is the shell model, the extension to the nucleus of the atomic orbital model. It is an old and natural idea; it goes back to the early 1930s, but at that time it met difficulties in both principle and practical implementation. In the first place, it was unclear how orbits could exist inside a nucleus, as it was presumed that nuclear collisions would disrupt any periodic motion. This is not so, however, because Pauli's exclusion principle prevents collisions if the final state is already occupied. Moreover, after the war the accumulated experimental data gave clear evidence of the existence of shells. Energy levels and quantum numbers of the orbits could be confidently assigned, and "magic numbers" appeared, indicating the filling of shells in the same way as in the chemical periodic system, in which the closure of shells is signaled by the occurrence of noble gases. The nuclear magic numbers are 2, 8, 20, 28, 50, 82, 126. When a nucleus contains a magic number of neutrons or protons, it shows particularly stable configurations. Maria Mayer (1906–1972), at Chicago (Figure 13.4) had studied a shell model in some detail, but it gave the wrong magic numbers. She tells how she complained of her problem to Fermi, who asked her, "Have you remembered to include the spin orbit coupling for nucleons?" Her answer was instantaneous: "No. And this will fix everything." Indeed, that was what happened.

In war-torn Germany, independently of Mayer, H. D. Jensen, P. Haxel, and H. Suess were doing something similar. Suess, of the same family as the Suess we met at page 37, was a geologist following the family tradition and was studying the abundance of elements. He was impressed by the indications of special stability for certain magic numbers and discussed the subject with his physicist friends. They ultimately reached the same conclusions as Mayer.

Although the shell models worked very well for nuclei having a number of nucleons close to a magic number, it did not work for nuclei with half-empty shells. The appropriate model for this case was devised by several physicists not long after the perfection of the shell model. Aage Bohr, the son of Niels, who as a young student had accompanied his father to Los Alamos, and B. Mottelson developed an original idea of James Rainwater of Columbia University, and considered collective motions of all nuclei in which the nuclear matter behaves more or less as a fluid. The shell model and the collective model represent limiting cases. In practice the two types of behavior coexist and influence each other. The combination of the two models is now very refined and accounts for a host of facts encompassing all nuclei.

Figure 13.4 Maria Mayer studying a chart of nuclei at the University of Chicago. This systematic study led her to create a shell model of the nucleus that is reminiscent of Bohr's atomic model and explains many nuclear regularities. (Photo University of Chicago.)

The discovery of the first transuranic elements, neptunium and plu-tonium, has had far-reaching consequences, including the practical ones we saw in Chapter 10. This, however, is not the end of the story; many more elements of ever-increasing atomic number have come into existence.

The method of preparation was always the same: the neutron addi-tion to a nucleus, followed by beta decay, which transforms the neutron into a proton. For instance, by irradiating Pu^{239} in a nuclear reactor, one obtains Pu^{240} and Pu^{242} by neutron capture, and these are transformed by beta decay into isotopes of americium of mass number 241 and 243, both of atomic number 95. By further bombardment of element 95 with neutrons, fol-lowed again by beta decay, one obtains element 96 (curium), and so on. To produce new transuranic elements there is a great advantage in using ex-tremely strong neutron sources, even atomic bombs, and some of the new elements have been found among the debris of atomic explosions. How-ever, one cannot proceed indefinitely both because the mean lives of the intermediate isotopes become too short and because other reactions compete favorably with the desired ones.

In the race to increase the atomic number, the neutron addition method reaches a point of diminishing returns. Physicists then use heavy ion bombardments. For instance, by using O^{16} as a projectile, it is possible to add eight protons at once. For this effect special accelerators are needed. The yield, however, is small, and in the end atoms are observed one by one by measuring their alpha particle emission or spontaneous fission.

The elements 95 (americium), 96 (curium), 97 (berkelium), 98 (californium), 99 (einsteinium), 100 (fermium), 101 (mendelevium), 102 (nobelium), and 103 (lawrencium) have thus been formed. Fourteen elements, starting with actinium, form a new rare earth family, and this is reflected in some of the names given to them; thus americium is a homolog of the rare earth europium. All transuranic elements are unstable either by alpha or beta emission or by spontaneous fission. Many of them have been prepared in Berkeley by groups of scientists, including G. T. Seaborg, A. Ghiorso, S. G. Thompson, and R. A. James, using either neutron bombardments or heavy ion bombardments.

More recently Soviet scientists, led chiefly by G. N. Flerov, have developed their own methods of the art and compete in the race toward higher atomic number. Numbers 105 and 106 are the last reached.

The Mössbauer Effect

This phenomenon straddles nuclear and solid-state physics and has found innumerable scientific applications. It was discovered in 1958 by Rudolf Mössbauer when he was still a student at the University of Munich. A free nucleus emitting a gamma ray recoils, and the recoil energy is subtracted from the transition energy. Usually the recoil energy is very small compared with the transition energy, and the recoil produces a small shift and widening of the spectral line emitted. In certain solids and under certain conditions, however, what recoils is not a single atom, but a small piece of the crystal of practically infinite mass. Spectral lines then become extremely sharp because the width is due to the uncertainty principle only. In favorable cases one reaches an energy definition of 1 part in 10^{12}. The sharpest line obtainable with an optical laser has about the same sharpness. The sharpness of the Mössbauer lines opens the possibility of observing phenomena otherwise completely inaccessible, such as the Zeeman effect of nuclear levels, some relativistic effects, many perturbations produced on the nucleus by the surrounding atoms in the crystalline lattice, chemical effects, and so on.

There are thousands of publications concerning the Mössbauer effect, and this avalanche is one more example of the interrelations between different fields of physics and of how a technique developed in one field may deeply influence another quite remote field. Other examples that come to mind are the Raman effect, nuclear magnetic induction, and, possibly the most spectacular, the use of tracers. These interconnections contrast sharply with the ever-increasing specialization prevailing in physics and emphasize the advantage or necessity of maintaining connections between different sciences. Chemists faced this problem long ago, and are perhaps quicker in incorporating "new weapons" in their research arsenal.

Superconductivity

After the preceding examples taken from theoretical, atomic, and nuclear physics, I would like to present examples of advances in solid-state physics. Unfortunately, they are remote from my personal technical knowledge; moreover, there is a great problem of selecting choice examples.

When I was a student, it was still possible to follow developments in all of physics, even if not in great detail. Fermi was up-to-date on any physics subject, and my friends F. Bloch, H. A. Bethe, and R. E. Peierls actively worked in both solid-state physics and nuclear physics. However, work in more than one field occurred almost exclusively among theoreticians, perhaps because theoretical technique is less differentiated than experimental technique (this is one of the great advantages and attraction of theory). There are recent examples of theoreticians such as R. Feynman, L. D. Landau, and C. N. Yang, who have made substantial contributions in very diverse fields, but these cases are infrequent; similar examples among experimentalists are even more rare.

If we consider the volume of research, the number of physicists employed, and the number of pages allocated in physics journals, solid-state physics is the largest field in physics today. This is natural, considering that it is also the field with the greatest industrial applications. It is not by chance that some of the best solid-state physics laboratories are financed directly by industry.

Since I speak very much as an outsider, I will limit myself to a couple of examples. The first is superconductivity. The phenomenon was discovered in 1911 by H. Kamerlingh Onnes as a fitting reward to his long efforts to reach ever-lower temperatures. For many years experimental and theoretical progress on the subject was moderate. The cryogenists at Leyden and elsewhere found transition temperatures, the effect of magnetic fields, behavior of specific heats, and other interesting but standard peculiarities of the phenomenon. In 1933 W. Meissner and R. Ochsenfeld in Berlin discovered a new aspect of superconductivity. Inside a superconductor the magnetic induction B is exactly zero. This remarkable property, as well as the vanishing of electric resistance, defines superconductivity. Meissner's discovery paved the way to the first macroscopic phenomenological theory of superconductivity, soon developed by the brothers F. and H. London (1935).

Their results showed precisely what would be required from a microscopic theory, but the passage from macroscopic to microscopic, that is, to a theory that could derive the properties of a superconductor from the atomic model, proved to be intractable for a long time. Only in 1957 did John Bardeen (b. 1908), Leon Cooper (b. 1930), and J. Robert Shrieffer (b. 1931) find the clue to the solution. The explanation is complicated and requires all the resources of quantum mechanics. It is remarkable that the mathematical scheme developed for dealing with superconductivity finds applications far removed from the original problem. For instance, nuclear physics seems to show examples of similar phenomena in nuclear matter. Thus the importance of the theory is not confined to the original problem, but extends to many others that deal with the behavior of systems containing many bodies. It provides a new way of thinking that is proving most fruitful.

Parallel to superconductivity there is another remarkable low-temperature phenomenon in liquid helium. An important part of the work on this subject originated in the USSR, where Piotr Kapitza (b. 1894) led in the experimental part and L. D. Landau (1908–1968) led in the theory. Below a certain critical temperature—2.18°K—He^4 develops a phase devoid of any viscosity, called *superfluid*. Superfluid helium shows some amazing behavior: It creeps along the walls of its container, literally escaping from the vessel that should confine it; it forms spectacular fountains; and so forth. Some of its properties had been guessed by Einstein in 1924, but nature proved much more imaginative even than Einstein.

Helium also has a rare isotope of mass 3, and helium 3 and 4 differ greatly in their macroscopic behavior at sufficiently low temperatures. Helium at low temperatures has thus become another large and surprising field of investigation. In the original theories the superfluidity of He^4 was connected to the fact that it obeys Bose statistics. However, when the rare isotope He^3, which obeys Fermi statistics, was accumulated in sufficient quantities, and cooled below 0.001°K, it also showed superfluidity! The theory had to be modified and extended and the ideas valid for superconductivity gave the cue.

Other Macroscopic Quantum Effects

Quantum mechanics is necessary to explain the very existence of atoms and thus, indirectly, of all structural properties of matter. It is, however, usually applied to systems so small as to be invisible. On the other hand, in solid-state physics there are several quantum mechanical effects that appear in macroscopic objects. One example is superconductivity and superfluidity; another, more typically macroscopic, is the quantization of magnetic flux passing through minute superconducting metallic tubes. It was observed independently at Stanford University and at Munich in 1961. Another example is the quantum mechanical tunneling through potential barriers. This had been well known for many years in the nuclear cases in which the potential barrier is due to the electrostatic field surrounding the nucleus. In the example to which I am referring, however, the potential barrier is a man-made oxide layer separating two conductors.

One of the most interesting cases of phenomena connected to barriers was predicted by B. Josephson in 1962 and realized the following year by P. W. Anderson and J. Rowell. When an oxide layer is sandwiched between two superconductors, an electric current passing through the sandwich is obtained. The current has two components: One is continuous, persisting even without an electromotive force just as it would in an ordinary superconductor; superimposed on it is a second alternating component, which appears only when a voltage difference V is established between the two superconductors. The frequency of this component is $\nu = 2eV/h$, independent of the materials used in forming the junction! The phenomenon gives a new method for measuring e/h once V is known. Conversely, if e/h is known, one can measure V. The phenomenon is so reliable and precise that it can be used to define the unit of voltage.

The study of materials that have electrical properties intermediate between metals and insulators—so-called semiconductors—led ultimately to the invention of the transistor. A necessary step in this achievement was the preparation of extremely pure germanium. Possibly no substance has been obtained in such quantities with so few impurities as germanium. With this material Bardeen (the same Bardeen who was later to find the clue to superconductivity), W. H. Brattain, and W. Shockley, at the Bell Laboratories, built the first transistors in 1948. The transistor is an extremely versatile constituent of all electronic circuits. Essentially it does the same things that a valve can do: amplify, rectify, oscillate, and so on; but it lacks a hot electron-emitting filament, and this has produced a true revolution in electronics. Without the transistor one could not build a modern computer or send a person to the moon. The transistor deeply affects our civilization because it has made computers possible and has altered our communication media.

At the Boundaries of Physics: Astrophysics, Biology

It is difficult to assign precise limits to physics. If physics is defined as what physicists do, it spills over into remote fields such as molecular biology, astrophysics, and geology.

Enormous strides have been made in astrophysics, astronomy, and cosmology mainly because of the technical advances in observing frequency ranges of electromagnetic radiation that were entirely inaccessible before the Second World War. Radioastronomy from the earth, and x-ray and ultraviolet astronomy from satellites have opened vistas entirely unsuspected before the war and have also directed the attention of ordinary telescope astronomers to inconspicuous optical objects of great interest. Quasars, pulsars, neutron stars, and black holes now fill the skies with unexpected cosmic dramas. Even the origin of the universe seems amenable to plausible hypotheses.

On a more modest scale, nuclear physics has given a satisfactory explanation of the origin of solar energy by detailed laboratory analysis of certain cycles of nuclear reactions that are sure to occur in the sun. One of these reactions is the synthesis of helium nuclei starting from protons. The cycle involved was discovered by H. A. Bethe in 1938. The final result is the combination of the four protons into a helium nucleus with a release of a large amount of energy; other nuclei have a vital role in the cycle as catalysts, and the electric charge is conserved by suitable positron emissions. As a theoretical discipline, nucleosynthesis is fairly developed and we have modest experiments pertaining to it on earth in both reactors and bombs.

The nuclear reactions occurring in the sun would be a most important energy source on earth if they could be produced not on an astronomical, but on a human scale and if they could be controlled. Studies to this effect are being carried out all over the world, with moderate success thus far. The central problem is to persuade the nuclei of a gaseous mixture to react with each other. This is attempted by electrical discharges or by other means such as heating with a laser beam. The old study of gas discharges has thus been transformed into plasma physics. Plasma is a mixture of atoms, electrons, and ions in a gaseous form. The immediate aim of plasma research is to obtain a plasma at sufficiently high pressure and temperature and to keep it together long enough to initiate a nuclear reaction between its constituents. These are usually ordinary hydrogen, deuterium, and tritium—the three isotopes of hydrogen of mass 1, 2, and 3. The energy liberated in the nuclear reactions that take place once the plasma is nuclearly ignited should suffice to keep it hot. The ultimate purpose is to extract the excess heat developed under controlled conditions, and to use it for practical purposes as any other source of thermal energy is used.

The difficulties facing this program are enormous, and so far success is still distant, but the possible advantages are also enormous because we would obtain an almost unlimited and "clean" energy source. In view of these extraordinary benefits and in the absence of obstacles of a fundamental nature, scientists and technologists, supported by their governments, persist in the quest. The chief immediate problem is that of containing the plasma. No material wall is likely to withstand the conditions prevailing in the plasma, and one has to contain it either by magnetic fields or by some as yet undiscovered trick.

On the other hand, explosive thermonuclear reactions are the basis of so-called thermonuclear bombs or hydrogen bombs, which, because of their destructive power, confront humanity with unprecedented dangers. In a hydrogen bomb an ordinary nuclear bomb ignites a mixture of light elements; and the power of the bomb is limited only by the amount of the reactants employed.

The new astronomy is discovering new and important gravitational effects. The study of gravitation was fundamentally stationary from Newton's time to the formulation of general relativity in 1916, when Einstein predicted three tiny effects difficult to observe. For many years general relativity remained a province occupied by a limited number of specialists somewhat outside the main stream of physics, but the situation is now changing. It has been possible to plan and start laboratory experiments on "gravitational radiation." Gravitational radiation is emitted in extremely small amounts, when masses are accelerated. Great cosmic events could radiate enough of it to make it detectable by terrestrial instruments.

Among other wonders of the skies, astronomers have seen objects with a mass comparable to that of the sun but with a radius of a few kilometers. Their density is of the order of 10^{14} g/cm^3, that is, of the same order of magnitude as that of nuclear matter and such that the volume of a drop of water would weigh as much as 500 large oil tankers. Their main constituent are neutrons. There are practically no electrons or protons. Since about 1934 Landau, Oppenheimer, and their pupils had postulated the existence and calculated the properties of neutron stars.

If a neutron star exceeds a certain limiting density, it collapses because the radiation pressure inside cannot compensate for the gravitational attraction. What then happens is not entirely clear. Extremely high densities are ultimately reached, and the gravitational field prevents any radiation from escaping the star. The star becomes a black hole, directly undetectable because no radiation can escape from it. Nevertheless, there are indirect ways of observing black holes, and there are some indications that they have in fact been observed.

On a much firmer foundation is the presence of a blackbody radiation corresponding to a present temperature of about 2°K that permeates the whole universe. This radiation is explained as a residue from a catastrophic explosion (big bang) that may have originated the universe 10^{10} years ago. After the big bang the universe has been expanding, and the blackbody radiation observed is the residue resulting from the transformation of the radiation present at the time of the explosion.

These are only a few of the revolutionary discoveries and theories that are now stirring astronomy. The skies have become populated with tremendous dynamic events totally unknown a few decades ago, and as-

Figure 13.5 The Crab Nebula, an extraordinary celestial body and a source of light, radio waves, and x-rays. In its center there exists a pulsating star, or pulsar, which is believed to be a neutron star rotating rapidly around its own axis. The nebula is the residue from the explosion of a supernova observed by Chinese astronomers in 1054. (Hale Observatories.)

trophysics is in a phase of rapid evolution comparable perhaps to that of nuclear physics in the 1930s. Astrophysics attracts many young and vigorous recruits, often originally trained as physicists. For some of them an attraction to the field is that no practical application is in sight for their endeavors (Figure 13.5).

It may be that molecular biology is the most significant and most far-reaching scientific conquest of the postwar era. For biology it is something like the discovery of quantum mechanics for physics, and we cannot foresee where molecular biology is going to lead us in the long run. Molecular biology certainly is not physics, but it is well anticipated by Corbino's words of 1929: "Better if we should realize not a simple superposition of techniques, but really obtain in the same brain a meld of the biological mentality with that created by the new physics." Indeed, not a few of the

Figure 13.6 (a) X-ray diffraction photograph of DNA taken by Rosalind Franklin in the laboratory of M. H. F. Wilkins. James Watson and Francis Crick found the clues to the double-helical structure of DNA in this photograph. (b) A space-filling model of the double helix of DNA. (Courtesy of M. H. F. Wilkins, Medical Research Council Biophysics Unit, Kings College, London.)   (a)

major leaders of the new discipline were trained and started their career as physicists. As a symbol of molecular biology, Figure 13.6 shows the famous double helix of Francis Crick and James Watson that has illuminated so many of the deepest problems of biology. To discover the double helix, Crick and Watson first had to know all the biological premises. These had to be combined with refined x-ray analyses of the structure, with tracer studies, and with a knowledge of the chemical bond intimately connected with quantum mechanics. Many different branches of science and technology thus had to converge on the object.

The Perplexed Scientist

I have briefly mentioned various currents of modern physics. They all have a high intellectual content, which is the main attraction to their practitioners, but we must not neglect their effect on the human condition. This is a hotly

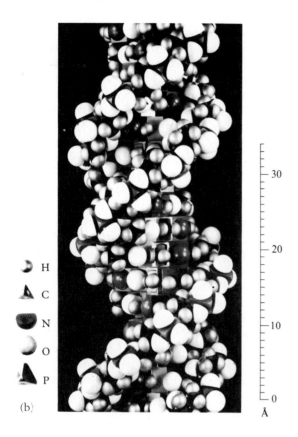

(b)

H

C

N

O

P

30

20

10

0

Å

debated argument that goes much beyond physics, though it is of prime importance to it. I will return shortly to the subject although I fully recognize that physicists as such have no special competence on the matter.

The influence of science on technology is a simpler theme, but not as simple as is sometimes believed. The transformation of a scientific discovery into a commercial product or new technology requires time, capital, ability, and favorable market and industrial surroundings. Above all, the personalities and driving motives of scientists seem to me different from those of inventors. I believe that the impulses that dominated Edison or Marconi were totally different from those that dominated J. J. Thomson or Hertz, not to speak of Einstein or Bohr. Many physicists have been at the root of important technological developments, but very few, if any, have participated in them in the industrial phase.

The modern way of feeding science into technology is also very different from that prevailing at the turn of the century. Now large teams tackle problems in wealthy laboratories, and the team's financial rewards do not depend entirely on its success. Team members are well schooled and also adaptable as workers in an organization. How different from individualist inventors like Edison or Marconi, who were only moderately educated and depended entirely on their own resources. Not only were patents and financial deals absolutely essential for them, but for many years they had to manage them personally. The interrelation of science and industry is most interesting, but beyond the scope of this book.

The change in organization of scientific endeavor that has occurred in this century is strikingly exemplified by a comparison of the conquest of air with that of space. The first was accomplished by inventors using their own idiosyncratic methods, the second by technologists using scientific methods. One of the most remarkable consequences is the small number of casualties in space conquest compared with the many who died in trying to fly. The reason is, in great part, that the pioneers of aviation had little understanding of the phenomena on which flight is based, whereas the planners behind the astronauts could calculate and experiment with almost everything necessary before actually venturing into space.

The conquest of space is an extreme example of tendencies characteristic of the industrial revolution, which began at the end of the seventeenth century largely through the ingenuity of rather uncultivated inventors, artisans, and entrepreneurs, and evolved into the second industrial revolution, dominated by research laboratories completely up-to-date on contemporary science with personnel fundamentally not very different from university scientists.

Let us turn now to the influence of science on the human condition. Many books have been written on the subject, and it would be presumptuous of me to try to discuss it in a few paragraphs. The problem is an old one, and we may refer to the words of Pierre Curie, quoted on page 42.

Certain points seem clear to me: First of all, science enhances human power. It also permits (at least approximately) anticipation of the consequences of certain courses of action. However, the process of decision at both the individual and the state level is dictated not by science but rather by obscure factors that I understand only dimly. They seem to me to be largely irrational, possibly dictated by behavioral forces, evolutionary drives, or subconscious demons. We thus see courses of action that to an outside observer appear totally irrational and destructive in their consequences. The armament race is an outstanding example. Because science enhances human power, it makes these foolish pursuits more and more dangerous, so much so that they may possibly imperil the survival of the species. The Luddite reaction advocated by some would be to put an end to science and thus prevent further increase in the human ability to do evil. It may be too late for such a solution even if it were feasible, but I believe it is not.

Furthermore, I find that to stifle curiosity runs contrary to deep human instincts. To deny the use of the brain because of fear of what one might find is unworthy of humanity. Dante Alighieri once put these words in Ulysses's mouth:

> Considerate la vostra semenza;
> Fatti non foste a viver come bruti
> Ma per seguir virtute e conoscenza.

> [Consider your progeny
> you were not created to live as brutes
> but to follow virtue and knowledge.]

(However, Ulysses's pursuit of his admonition brought him and his companions to catastrophe.)

The optimistic and perhaps Panglossian point of view is that in due course people will succeed in applying their rational powers to decision making. I cannot say that I am too sanguine about this optimistic view, chiefly because the time required to change mental attitudes may not be available.

What are we to do? And, especially, what should scientists do? We should try to constantly remind the public of the consequences of certain courses of action. This should be done honestly and intelligently.

I am aware that, unfortunately, objectivity in these subjects is elusive and that politics, economics, and social behavior are not amenable to the same type of treatment as science. The complexity of the issues, the impossibility of treating them separately, the nature of the data involved, qualitatively different from those dealt with in the infinitely simpler physical sciences, are only a few of the obstacles to the applications of scientific methods to human social problems.

While the scientist has the specialized knowledge of his discipline, on other subjects he is pretty much prey of the same dark forces as is anybody else. His training and education may help him to overcome some of his irrational urges, but the idea that the objective, cool scientist is above the crowd is fallacious. This should be recognized by the scientists and by the public at large. Scientists are not priests of a magic religion.

The duty of informing his fellow man is paramount, but the responsibility for the use made of his discoveries cannot honestly be pushed onto the scientist. He does not have the power to determine what use will be made of his findings, and one cannot separate power from responsibility. Furthermore, practically all discoveries can be applied to a diversity of purposes, some of which appear to us good, and some evil. This has been known since the beginning of civilization—a symbol of the dilemma is the use of steel for the making of swords or plowshares. In modern times penicillin and DDT have been properly hailed as boons to mankind, only to find out that their indiscriminate use may have dire indirect consequences. Nuclear energy may be a great help for mankind, or its undoing. The choice is not in the hands of the scientists.

Chapter 14

Conclusions

I would like to bring this book to an end with some general conclusions. First of all, no portrait of the "physicist" emerges from our study. There is, rather, a great variety of personages, not surprising considering the great diversity of contributions necessary to the advancement of physics. As extreme cases, we may think of E. O. Lawrence and P. A. M. Dirac; a comparison reveals contrasting personalities, yet both have been important to the progress of physics. Two great stars—Rutherford and Einstein—would have had difficulty in carrying on a serious scientific discussion with each other. Their interests, culture, and scientific imagination differed too much. It would have been difficult to make Rutherford feel the importance of the invariance of Maxwell's equations under the Lorentz transformation or to interest Einstein in the technical details necessary to pin down nuclear disintegration.

If we are determined to single out traits common to important physicists, we find a great eagerness and capacity for work, stamina, optimism, and scientific imagination. But these are the qualities needed for success in almost any profession from athlete to poet, from banker to general. Intellectual qualities such as analytical ability and theoretical imagination are exceptional in some, but modest in others who excelled nevertheless. Some physicists are universalists, some narrow specialists. A person such as H. A. Lorentz would have been eminent in almost any endeavor, including a diplomatic or a business career. However, what would Pauli have done other than theoretical physics? The conclusion is that all roads *may* lead to Rome, not, of course, that all roads *do* lead to Rome.

Although luck plays a role in scientific findings, there are too many cases of simultaneous discoveries to attribute a great deal to chance. Rather, the historical moment in which one works is more important. There are fat periods and lean periods. Moreover, there seems to be a certain internal logic in the development of a scientific discipline, and the personality cult should not be carried too far. Think of the simultaneous discovery of three forms of quantum mechanics. All in all, we have the sobering impression that even if one of the greatest physicists had not been born, fifty years later physics would nevertheless be at the same point.

Future Trends

Let us turn now to the present state of physics. In the first place, we must distinguish physics from its applications. Applications can be economically and socially much more important than pure physics, but I am concerned here with physics as the exploration of natural phenomena. This comprises the description, classification, and interrelation of different phenomena, and their synthesis into more or less comprehensive theories.

It seems to me that particle physics is the one area with the greatest unsolved intellectual problems. Some of them are almost a century old, and physicists tend to forget them because there is no inkling of their solution. For instance, nobody knows why the electric charge is always an integral multiple of that of the electron. Here the word *why* has to be understood as the connection of a fact with another fact. The "reason" is mostly simply a logical or mathematical argument that shows that if *a* is true, *b* follows. The quantization of the electric charge has not been connected to any other experimental fact; in this sense it is a primary empirical datum.

Other problems basic to elementary particle physics that are simple to formulate and yet of unknown solution are the conservation of nucleons and the similar conservation of leptons. The conservation of nucleons simply means that, in a closed system or in the universe, the sum of the number of protons plus the number of neutrons is constant. In the reckoning, antiprotons and antineutrons must be, of course, counted with the minus sign. The conservation of nucleons is well substantiated because it is known that the mean life of protons is longer than 10^{22} years, that is, 10^{12} times the age of the universe. However, the source of this enormous stability of protons, which is at the basis of the conservation of nucleons, remains unknown. A similar conservation principle, also of unknown origin, obtains for leptons. Furthermore, electrons and muons each have corresponding neutrinos. The two kinds of neutrinos are distinct because a muonic neutrino colliding with a neutron can produce a muon and a proton, but not an electron and a proton, while an electronic neutrino colliding with a neutron can produce an electron and a proton, but not a muon and a proton. What makes the difference between a muonic and an electronic neutrino? Electrons and muons both seem to be exactly described by Dirac's equation except for the mass that enters as a parameter in the equation. Why should nature choose to make two particles that differ only in mass but are otherwise identical? All the above are examples of unexplained or unconnected facts. Recently, a third lepton, τ, has been discovered. Is this only the beginning of a finite (or infinite?) series of leptons?

Contrasting examples of connected or "explained" phenomena are the conservation of momentum or angular momentum. It is possible to show that they are mathematical consequences of the isotropy and homogeneity of space, and in this sense one thing explains the other. Great progress is achieved when we uncover deep connections that are aesthetically satisfying and help to reduce the number of independent "postulates." I have put the word *postulates* in quotation marks because the word has a meaning in physics different from that usually intended in mathematics, where a postulate is an essential premise of a train of reasoning and is largely arbitrary. In physics postulate means, in substance, an experimental fact that has been observed many times without exception. Mathematical postulates are much more arbitrary and are essentially a creation of the human mind. Physical postulates are based on observation and extrapolation and subordinately on aesthetic criteria. One can develop a non-Euclidean geometry, indeed several of them, but a physics that would disagree with experiment is not physics.

Experiments, however, are always imprecise and are affected by errors; hence, in interpreting them there is always a process of abstraction and idealization. Physicists have learned historically how to proceed and are accustomed to using successive approximations and to limiting the field of applicability of their theories. The most famous example is Newtonian mechanics and its improvements, relativity and quantum mechanics. Whatever the "postulates" of physics are, they do not create terrible doubts in the minds of most practicing physicists. This does not mean either that a down-to-earth experimentalist ignores the epistemological problems of his science or that even the most critical theorist is stopped cold in his calculations by epistemological doubts. People have learned to live with them; even Bohr jokingly said that the opposite of a shallow truth is untrue but the opposite of a deep truth is also true ...

Let us return, however, to the simple mysteries of particle physics mentioned above. There are some beginnings of answers and some attempts at major syntheses of diverse theories. The most comprehensive was initiated by Steven Weinberg and Abdus Salam, independently, in 1967–1968 (p. 269). They tried to unify weak interactions and electromagnetism. One hundred years earlier the grandiose Maxwellian synthesis had finally succeeded in unifying electricity and magnetism; the present attempt could be considered in the same light. So far experiments seem to fit the theory, and new discoveries of weak neutral currents and charmed quarks are to be considered as evidence of success of this approach. Some important verifications will be possible at higher energies than are available now: There should be some particles called intermediate bosons that are analogous to light quanta inasmuch as they should be the quanta of the weak interactions. They are expected to be very heavy and thus obtainable only at high energy. Their discovery would be a strong corroboration of this trend of ideas.

In other branches of physics there are also most interesting problems of more practical importance than those offered by elementary particles, but we can confidently believe that their solution is implicitly contained in Schrödinger's equation. Of course, it is contained in the same sense that marble blocks contain all statues, and the problem is only to take away the excess material. Michelangelo's sonnet expressing this thought was intended not as a joke but as a philosophical truth.

There are also problems of fundamental theoretical technique pertaining to multibody systems that require solution. They would greatly advance our understanding of liquids, nuclear matter, and other aggregates, and surprisingly, perhaps, even make inroads into particle physics. They are important, but I would not expect from multibody theories the startling revelations we have had from particle physics—a new type of force or the nonconservation of parity. However, I could be wrong. In my opinion the possibility of intellectual discovery points to particle physics. This is a subjective opinion, and since it might have serious consequences for scientific policy, I want to stress that this opinion is not unanimous.

In the field of particle physics, which necessarily implies high energies, costs and experimental complications have grown alarmingly. As a consequence, fewer and fewer experiments can be performed, and they last longer and longer. When experiments were cheap, imaginative physicists could develop their ideas and perform many experiments. The great Faraday listed more than 10,000 experiments in his laboratory notebooks. In high-energy physics, when an experiment costs millions of dollars, lasts five years, and requires dozens of experimenters, one can perform only a few. The experimenters inevitably face problems in a different spirit. Committees often decide what to do, and it is hard to trust the imagination of a committee, which is less adventurous and wants to be reasonably sure that the experiment will give some tangible result. Physicists are thus led to try safe experiments that are obvious and acquire their interest from the unusual energy region at which they are performed, being otherwise rather simpleminded. I once compared such experiments to modern warfare between naval vessels in which the range of the battleship guns may be decisive; at the time of Lord Nelson, by contrast, the admiral had to have uncanny tactical intuitions. However, "battleship experiments" have given some very great surprises and have proved vital to progress.

If the type of experimental information desired is fairly obvious, the difficulties in obtaining it are principally technical; hence physicists must devote most of their efforts to the development of instruments and techniques, including programs for analyzing results by computer. Once an instrument is developed, it is used to perform all the experiments for which it is suitable. All this has its logic and usually leads to important results, but it requires physicists somewhat different from the traditional ones.

With a pun of dubious taste, I could say that it is no longer sufficient to be a Rutherford, but one must be a Ruther-Ford, meaning that the physicist must have at least some of the qualities of an industrialist and of a businessman. In the past many physicists were pushed to science by a certain coolness or shyness in human relations and a greater attraction to things; (the Curies are an example). Today, interest in people may be a paramount requirement in leading positions (for example, E. O. Lawrence). All this is not completely novel, but the role of the isolated investigator, such as Faraday or Röntgen, seems destined to wane.

Another tendency pervades all sciences, including physics: specialization. Both the physics literature and the number of physicists have multiplied enormously, with the consequence of ever-increasing specialization. It is an inevitable evil, which we must accept. Those who have witnessed the smaller, simpler, and more unified physics may feel nostalgic about it, but the trend is irreversible.

The trend is well illustrated by comparing Figure 3.5 which represents Rutherford's first studies of nuclear scattering and disintegration, with Figure 11.11 representing the Fermi lab. The experiments done with the two facilities are in principle the same. Moreover, this problem is not limited to experiment alone, as shown by an example jokingly reported by J. D. Jackson, when he compared a modern formula (Figure 14.1b) with Rutherford's formula as given in the *Philosophical Magazine* of 1911 (Figure 14.1a). The former is typical of much modern work, though the latter is more famous.

This increasing complication may even determine the limits of physics. Up to now we have found successive layers on a descending scale of

$$y = \frac{Qdm}{2\pi r^3 \sin\phi \cdot d\phi} = \frac{ntb^2 \cdot Q \cdot \operatorname{cosec}^4 \phi/2}{16r^3} \cdot \cdot \cdot \ (5)$$

(a)

Figure 14.1 (a) "My thanks are due to Mr. W. Kay for his assistance in counting and in all the experimental work." (Profesor Ernest Rutherford, April 1919.)
(b) "We wish to thank Professor J. Steinberger who participated in the design and in the earlier part of this experiment, and Professors W. Paul, P. Preiswerk and K. Faissner for support and encouragement. We acknowledge the assistance of Mr. J. Daub and Dr. P. Zanella who made the measurement of the events on Luciole possible. Dr. L. Caneschi has helped in the running of the experiment. The detection apparatus was built with the help of the Messrs. F. Blythe, K. Bussmann, J. M. Fillot and G. Maratori. Finally we would like to thank Dr. G. Petrucci, the CPS staff and especially Dr. L. Hoffmann for the setting up and operation of the slow ejected proton beam." (A. Bohm, P. Darriulat, C. Grosso, V. Kaftarev, K. Kleinknecht, H. K. Lynch, C. Rubbia, H. Ticho, and K. Tittel, May 30, 1968.)

size and on an increasing scale of energy: atoms, nuclei, subnuclear particles such as pions, quarks. It is conceivable that the quarks may allow a closed self-consistent theory, but I cannot suppress the sneaking suspicion that the chain may not end there. If so, more money and more effort will be needed for the next step, and the process may diverge so as always to leave us with a dangling end. The limit will then be the amount of effort that people can or are willing to devote to the search, without any guarantee that it will end.

I will close with the remark that the trend toward abstraction and removal from immediate sensory experience is not limited to physics; on the contrary, it seems almost a characteristic of modern thought and art (Figure 14.2). In the Curies' time contemporary painters described reality, although they personalized it. At the time of the development of quantum mechanics, reality is already remarkably deformed in the mind of the artist. Picasso painted women with two faces, although not one corpuscular and the other wavelike. We now have totally abstract paintings and in physics formulae of a deep but not immediate meaning.

The Innards of Physics

I cannot escape the feeling that science resembles a living organism in its structure and evolution. At the basis of the analogy is the complexity of the scientific system. Let us consider physics only, for the sake of definite-

$$
\begin{aligned}
\frac{d\sigma}{dt} = \frac{1}{4\pi(s-m^2)^2}\Bigg[& \frac{|\sin\theta_t|^2(1+\cos2\varphi_\gamma)|t-(Y+m)^2||t|^{-1}|\gamma_{\frac{1}{2}\frac{1}{2}}{}^K|^2(s/s_0)^{2(\alpha_K-1)}\alpha_K{}^2}{|\Gamma(\alpha+1)\sin\frac{1}{2}\pi\alpha_K|^2} + (1-\cos2\varphi_\gamma) \\
& \times|\sin\theta_t|^2(t-m_K{}^2)^2|t-(m-Y)^2|\left(\frac{\alpha_c{}^2|t|^{-1}|\gamma_{\frac{1}{2}\frac{1}{2}}{}^c|^2(s/s_0)^{2(\alpha_c-1)}}{|\Gamma(\alpha_c+1)\sin\frac{1}{2}\pi\alpha_c|^2} + \frac{\alpha_V{}^2|t|^{-1}|\gamma_{\frac{1}{2}\frac{1}{2}}{}^V|^2(s/s_0)^{2(\alpha_V-1)}}{|\Gamma(\alpha_V+1)\cos\frac{1}{2}\pi\alpha_V|^2}\right. \\
& + \frac{\alpha_T{}^2|t|^{-1}|\gamma_{\frac{1}{2}\frac{1}{2}}{}^T|^2(s/s_0)^{2(\alpha_T-1)}}{|\Gamma(\alpha_T+1)\sin\frac{1}{2}\pi\alpha_T|^2} + \frac{2\alpha_c\alpha_V\sin\frac{1}{2}\pi(\alpha_V-\alpha_c)\gamma_{\frac{1}{2}\frac{1}{2}}{}^c\gamma_{\frac{1}{2}\frac{1}{2}}{}^V(s/s_0)^{\alpha_c+\alpha_V-2}}{|\Gamma(\alpha_c+1)\Gamma(\alpha_V+1)\sin\frac{1}{2}\pi\alpha_c\cos\frac{1}{2}\pi\alpha_V|} \\
& \left. + \frac{2\alpha_c\alpha_T\gamma_{\frac{1}{2}\frac{1}{2}}{}^c\gamma_{\frac{1}{2}\frac{1}{2}}{}^T\cos\frac{1}{2}\pi(\alpha_T-\alpha_c)(s/s_0)^{\alpha_c+\alpha_T-2}}{|\Gamma(\alpha_c+1)\Gamma(\alpha_T+1)\sin\frac{1}{2}\pi\alpha_c\sin\frac{1}{2}\pi\alpha_T|}\right) + (1+\cos^2\theta_t-\cos2\varphi_\gamma\sin^2\theta_t) \\
& \times \frac{\alpha_K{}^4|t-(m+Y)^2|t^{-2}|\gamma_{\frac{1}{2}-\frac{1}{2}}{}^K|^2(s/s_0)^{2(\alpha_K-2)}}{|\Gamma(\alpha_K+1)\sin\frac{1}{2}\pi\alpha_K|^2} + (1+\cos^2\theta_t+\cos2\varphi_\gamma\sin^2\theta_t)(t-m_K{}^2)^2 \\
& \times|t-(m-Y)|^2\left(\frac{|\gamma_{\frac{1}{2}-\frac{1}{2}}{}^c|^2(s/s_0)^{2(\alpha_c-1)}\alpha_c{}^2|t|^{-2}}{|\Gamma(\alpha_c+1)\sin\frac{1}{2}\pi\alpha_c|} + \frac{\alpha_V{}^4|\gamma_{\frac{1}{2}-\frac{1}{2}}{}^V|^2(s/s_0)^{2(\alpha_V-1)}}{|\Gamma(\alpha_V+1)\cos\frac{1}{2}\pi\alpha_V|} + \frac{\alpha_T{}^2|\gamma_{\frac{1}{2}-\frac{1}{2}}{}^T|^2(s/s_0)^{2(\alpha_T-1)}}{|\Gamma(\alpha_T+1)\sin\frac{1}{2}\pi\alpha_T|^2} + 2\sin\frac{1}{2}\pi\right. \\
& \times(\alpha_V-\alpha_c)|t|^{-1}\frac{\gamma_{\frac{1}{2}-\frac{1}{2}}{}^c\gamma_{\frac{1}{2}-\frac{1}{2}}{}^V(s/s_0)^{\alpha_c+\alpha_V-2}}{|\Gamma(\alpha_c+1)\Gamma(\alpha_V+1)\sin\frac{1}{2}\pi\alpha_c\cos\frac{1}{2}\pi\alpha_V|} + \left.\frac{2\cos\frac{1}{2}\pi(\alpha_T-\alpha_c)\gamma_{\frac{1}{2}-\frac{1}{2}}{}^c\gamma_{\frac{1}{2}-\frac{1}{2}}{}^T(s/s_0)^{\alpha_c+\alpha_T-2}}{|\Gamma(\alpha_c+1)\Gamma(\alpha_T+1)\sin\frac{1}{2}\pi\alpha_c\cos\frac{1}{2}\pi\alpha_T|}\right) + 4\cos\theta_t(t-m_K{}^2) \\
& \times[(t-(m+Y)^2][t-(m-Y)^2]|^{1/2}\gamma_{\frac{1}{2}-\frac{1}{2}}{}^K(s/s_0)^{\alpha_K-1}\alpha_K{}^2\left(\frac{t^{-2}\gamma_{\frac{1}{2}-\frac{1}{2}}{}^c\alpha_c\cos\frac{1}{2}\pi(\alpha_K-\alpha_c)(s/s_0)^{\alpha_c-1}}{|\sin\frac{1}{2}\pi\alpha_c\Gamma(\alpha_c+1)\Gamma(\alpha_K+1)\sin\frac{1}{2}\pi\alpha_K|}\right. \\
& + \left.\frac{t^{-1}\sin\frac{1}{2}\pi(\alpha_V-\alpha_c)\alpha_V\gamma_{\frac{1}{2}-\frac{1}{2}}{}^V(s/s_0)^{\alpha_V-1}}{|\Gamma(\alpha_K+1)\Gamma(\alpha_V+1)\sin\frac{1}{2}\pi\alpha_K\cos\frac{1}{2}\pi\alpha_V|} + \frac{t^{-1}\alpha_T\gamma_{\frac{1}{2}-\frac{1}{2}}{}^T\cos\frac{1}{2}\pi(\alpha_K-\alpha_T)(s/s_0)^{\alpha_T-1}}{|\Gamma(\alpha_T+1)\Gamma(\alpha_K+1)\sin\frac{1}{2}\pi\alpha_K\sin\frac{1}{2}\pi\alpha_V|}\right)\Bigg]. \quad (75)
\end{aligned}
$$

(b)

Figure 14.2 A tendency toward abstraction in the arts parallels that which seems to dominate the development of physics in the twentieth century. (a) *Grey Tree* by P. Mondrian (1872–1944), in which the form of the tree is clearly recognizable, although it is not a figurative drawing. (b) Mondrian's *Apple Tree in Bloom,* dated 1912. The form of the tree has disappeared, giving way to a collection of geometric curves, which nevertheless seem to recall the earlier form. (Collection Haags Gemeentemuseum, The Hague, The Netherlands.)

(a)

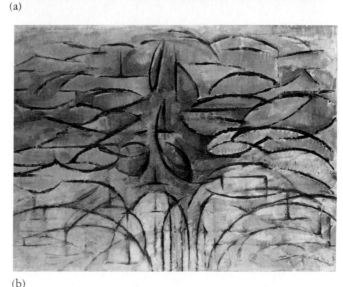

(b)

ness. Experiment and theory are its fundamental components, and there is a steady give and take between them. However, experiment requires instruments and techniques on the material side and goals and ideas on the intellectual side. The technical needs give the tie to technology, which in turn is steadily enriched by science. The need of goals give the tie to theory. Theory in turn without a subject matter and a continuous check on its veracity furnished by experiment cannot survive as a fruitful endeavor. Furthermore, theory too needs its specific tools—mathematics, and mathemat-

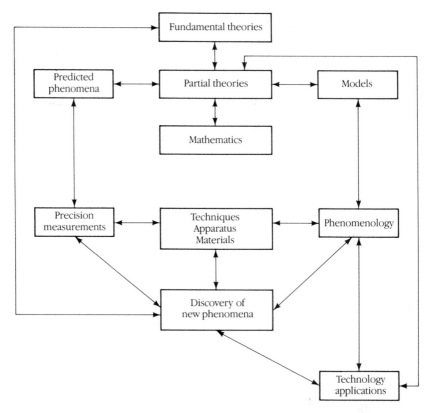

Figure 14.3 The different parts forming "physics" resemble in their interdependence the organs of a living being. They show a very complex system of interrelated activities, and they evolve in time.

ics for a long time has drawn its inspiration from physics, while occasionally abstract mathematical theories created without any regard to applications have surprisingly found their use in physics. In this picture we must fit also the occasional chance discovery, which plays a role similar to that of mutation in biology. They introduce new departures that may evolve into entirely new chapters of science. These in turn have to be connected with the existing part of the organism and thus induce changes in it.

I have tried to symbolically represent some aspects of these interrelations in Figure 14.3. Each of the squares makes me think of an organ in a living being, whose amputation could cripple or kill the whole organism. The connections remind me of the reaction of one organ on the other. But enough allegory. We could substantiate each square and each connection with examples drawn from real events, and this book could furnish a great number of them.

Given the complexity of this scheme it is little wonder that such different types of persons can contribute mightily to the development of physics. Each interaction is better carried out by a certain personality, and thus there is the possibility of fruitful work for all types of people.

The resemblance to a living thing also covers evolution. Not only the subject matter, but the philosophy of physics too, changes with time, and there is every reason to think that it will continue to do so, even at a profound level. I do not believe Galileo, Newton, and Einstein have been the last of their ilk.

Appendix 1

Stefan's Law; Wien's Law

Stefan's law $u(T) = aT^4$ follows simply from thermodynamics and the relation $p = u(T)/3$ between $u(T)$ and the radiation pressure p. This relation is a consequence of Maxwell's equations for the electromagnetic field.

The first principle of thermodynamics gives

$$dQ = dU + p\, dV$$

where $U = u(T)\, V$ is the total internal energy of the radiation, V the volume, and Q the heat supplied.

The second principle states that $dS = dQ/T$ is an exact differential. It follows by combining the first and second principle that $dS = [u\, dV + V\, du + (u/3)dV]/T$ is a total differential in T and V. Hence

$$\frac{\partial^2 S}{\partial T \partial V} = \frac{1}{T}\frac{d\, u(T)}{dT} = \frac{4}{3}\frac{d(u/T)}{dT}$$

From this we obtain at once $du/dT = 4u/T$ or

$$u(T) = aT^4$$

Wien, by a detailed thermodynamical argument in which he considered the compression of black radiation by a moving piston perfectly reflecting, obtained

$$u(\nu,T) = \nu^3 f(\nu/T)$$

Stefan's law follows from Wien's law by integration in $d\nu$. Taking the derivative of $u(\nu,T)$ with respect to ν and finding where it is zero, we obtain the frequency at which $u(\nu,T)$ is maximum at a given temperature. Using Wien's law, we find $3f(\nu/T) = (\nu/T)f'(\nu/T)$. The maximum corresponds to a certain numerical value of ν/T. Or if one chooses λ and T as variables and calculates $u(\lambda,T)d\lambda$ one finds similarly a value of λT. This property is called sometimes the displacement law. Using Planck's expression for $u(\nu,T)$ one finds numerically $\nu_m/T = 5.88 \times 10^{10}$ (ν in cycles/sec, T in degrees K).

Planck's Hunt for the Blackbody Radiation Formula

Planck first established the relation between the radiation density and the average energy $\langle E \rangle$ of the oscillators forming the walls of the blackbody. By thermodynamics he obtained

$$u(v,T) = \frac{8\pi v^2}{c^3} \langle E \rangle \tag{1}$$

He also considered the average entropy of an oscillator and showed that it obeys the equation

$$dS = \frac{d\langle E \rangle}{T} \tag{2}$$

If we want to obtain for $u(v,T)$ the expression put forward by Wien

$$u(v,T) = A v^3 \exp(-\beta v/T) \tag{3}$$

which seemed to be supported by experiment, we can write using Equations (1), (2), and (3):

$$\frac{dS}{d\langle E \rangle} = \frac{1}{T} = -\frac{1}{\beta v} \log \frac{8\pi \langle E \rangle}{A c^3 v} \tag{4}$$

and by making the second derivative

$$\frac{d^2 S}{d\langle E \rangle^2} = -\frac{1}{\beta v \langle E \rangle} \tag{5}$$

We obtain these results by solving Equation (3) for $1/T$ and using Equations (1) and (2).

On the other hand, if we want to obtain the Rayleigh-Jeans formula, we have $\langle E \rangle = kT$ and

$$\frac{dS}{d\langle E\rangle} = \frac{1}{T} = \frac{k}{\langle E\rangle} \tag{6}$$

$$\frac{d^2S}{d\langle E\rangle^2} = -\frac{k}{\langle E\rangle^2} \tag{7}$$

Experiment showed that neither formula was correct, but that each represented limiting cases. We thus want an interpolation that for $\langle E\rangle \gg k\beta\nu$ coincides with Equation (7) and for $\langle E\rangle \ll k\beta\nu$ coincides with Equation (5). It is then tempting to try for $d^2S/d\langle E\rangle^2$ the interpolation formula that has the proper limiting values

$$\frac{d^2S}{d\langle E\rangle^2} = \frac{-1}{\beta\nu\langle E\rangle + \langle E\rangle^2/k} \tag{8}$$

Assuming this formula to be correct, integration in $d\langle E\rangle$ gives

$$\frac{dS}{d\langle E\rangle} = -\frac{1}{\beta\nu}\log\frac{k\beta\langle E\rangle}{1 + k\beta\langle E\rangle} = \frac{1}{T} \tag{9}$$

where we have used the boundary condition that for $T \to \infty$ also $\langle E\rangle$ goes to infinity. We can now solve Eq. 9 for $\langle E\rangle$ and by inserting the result into Equation (1) and calling $h/k = \beta$ obtain:

$$\langle E\rangle = \frac{k\beta\nu}{e^{\beta\nu/T} - 1} \tag{10}$$

which inserted in Equation (1) gives

$$u(\nu,T) = \frac{8\pi\nu^2}{c^3} \cdot \frac{h\nu}{e^{h\nu/kT} - 1} \tag{11}$$

Appendix 3

Einstein's Heuristic Argument for Postulating the Existence of Light Quanta

Define an entropy density $\varphi(\nu,T)d\nu$ for blackbody radiation similar to the energy density $u(\nu,T)d\nu$. By thermodynamics one can show that

$$\frac{d\varphi(\nu,T)}{du(\nu,T)} = \frac{1}{T} \tag{1}$$

Assuming Wien's radiation formula

$$u(\nu,T) = A\nu^3 e^{-h\nu/kT} \tag{2}$$

and solving for $1/T$, one finds

$$\frac{1}{T} = -\frac{k}{h\nu} \log \frac{u}{A\nu^3} = \frac{d\varphi}{du} \tag{3}$$

and integrating

$$\varphi(u,\nu) = -\frac{ku}{h\nu}\left[\log \frac{u}{A\nu^3} - 1\right]$$

These remarks correspond to Planck's reasonings (App. 2) and were known before Einstein. Consider now a change of volume of a blackbody, leaving the total energy constant. This requires that $u(\nu,T)V = U(\nu,T)$ remain constant. The entropy in the frequency interval $d\nu$ is

$$dS = \varphi(\nu,T)Vd\nu = -\frac{kU}{h\nu}\left[\log \frac{U}{AV\nu^3} - 1\right]d\nu \tag{4}$$

For a finite volume change, keeping U constant, we have:

$$d(S - S_o) = \frac{kU}{h\nu} \log \frac{V}{V_o} d\nu \tag{5}$$

For a perfect monatomic gas of n atoms thermodynamics gives for the entropy change by an expansion at constant energy:

$$S - S_o = nk \log \frac{V}{V_o} \qquad (6)$$

This last formula is easily connected to Boltzmann's expression

$$S = k \log W$$

because for one atom the probability of being in a partial volume V of V_o is V/V_0. For n independent atoms it is then $(V/V_o)^n$ from which Equation (6) follows immediately.

Equations (5) and (6) become the same if one assumes

$$U \, dv = nhv$$

That means that the energy in frequency interval dv is subdivided in n quanta of magnitude $\varepsilon = hv$.

Appendix 4

Brownian Motion

A particle of mass m and radius a floats in a liquid at temperature T. The molecules of the liquid impinge on the particle and push it sometimes to the left sometimes to the right. The particle is large enough to have, on average, its motion damped by the viscosity of the liquid. The motion of the particle is then essentially a random walk, with steps to the right equally as probable as steps to the left.

Calling $\langle x^2 \rangle$ the average square displacement after N steps of length λ, one has according to elementary statistics

$$\langle x^2 \rangle = \lambda^2 N \tag{1}$$

If the average speed of the particle is v, the number of steps in time t is

$$N = vt/\lambda \tag{2}$$

The speed v may be obtained by the equipartition of energy, which says that

$$\tfrac{1}{2}mv^2 = \tfrac{3}{2}kT \tag{3}$$

The length λ may be estimated by assuming that it is such that the viscosity force acting for a length λ does a work equal to the kinetic energy of the particle. If the particle started in one direction after the displacement λ, it has lost its speed and may turn around. Stokes's law gives for the viscosity force the expression

$$F = 6\pi a \eta v$$

where η is the viscosity coefficient of the liquid. It follows that

$$6\pi a \eta v \lambda = \tfrac{3}{2}kT \qquad \text{or} \qquad \lambda = \frac{kT}{4\pi\eta av}$$

Using this expression and Equations (1), (2), (3), one has, very crudely,

$$\langle x^2 \rangle = \frac{kT}{4\,\pi a \eta}\, t$$

Einstein's calculation gives:

$$\langle x^2 \rangle = \frac{kT}{3\,\pi a \eta}\, t$$

Appendix 5

Blackbody Energy Fluctuations According to Einstein

Consider an isolated system of energy E and divide it into two parts. They have energy E_1 and E_2 with $E_1 + E_2 = E$ and similarly entropies $S_1 + S_2 = S$. When everything is in equilibrium, the same quantities are denoted by E_1^0, E_2^0, S_1^0, S_2^0. We define $E_1 - E_1^0 = \varepsilon = E_2^0 - E_2$. We also assume that part 1 is small compared with part 2 of the system.

We develop the entropy in a power series in ε

$$S = S_1 + S_2 = S_1^0 + S_2^0 + \left(\frac{\partial S_1}{\partial E_1} - \frac{\partial S_2}{\partial E_2} \right) \varepsilon + \left(\frac{\partial^2 S_1}{\partial E_1^2} + \frac{\partial^2 S_2}{\partial E_2^2} \right) \varepsilon^2 + \cdots$$

In equilibrium conditions the entropy is a maximum and the coefficient of ε vanishes. This condition means that

$$\frac{\partial S_1}{\partial E_1} = \frac{\partial S_2}{\partial E_2} = \text{const.} = \frac{1}{T_1} = \frac{1}{T_2}$$

and by definition of temperature $\left(\dfrac{\partial S}{\partial E} = \dfrac{1}{T} \right)$ it follows that the two parts are at the same temperature.

The next term in ε^2 gives

$$\Delta S_1 = S_1 - S_1^0 = \frac{\partial^2 S}{\partial E_1^2} \varepsilon^2 = \frac{d \frac{1}{T_1}}{dE_1} \varepsilon^2 = -\frac{1}{T^2} \frac{dT}{dE_1} \varepsilon^2 = -\frac{\varepsilon^2}{T^2 C_{v,1}}$$

We have here used the fact that the part 2 is very large compared with part 1 and thus has a much larger heat capacity. Furthermore, by definition $dE_1/dT_1 = 1/C_{r,1}$ where $C_{r,1}$ is the heat capacity of part 1. At the end we have suppressed the index 1 because it is superfluous.

Boltzmann's relation between entropy and probability gives as probability for an energy fluctuation of magnitude ε

$$W(\varepsilon) = e^{-\Delta S/k} = e^{-\varepsilon^2/kT^2 C_v}$$

and for $\langle \varepsilon^2 \rangle$ we find:

$$\langle \varepsilon^2 \rangle = \frac{\int_{-\infty}^{\infty} \varepsilon^2 W(\varepsilon)\, d\varepsilon}{\int_{-\infty}^{\infty} W(\varepsilon)\, d\varepsilon}$$

The integrals are standard and performing them we find

$$\langle \varepsilon^2 \rangle = kT^2 C_v \tag{1}$$

This result is to be found in Einstein's papers of 1903. We now apply it to blackbody radiation. The energy E in a volume V and frequency interval $\Delta \nu$ is given by Planck's formula. We calculate the heat capacity C_v by making the derivative with respect to T and insert the result in Equation (1) obtaining

$$\langle \varepsilon^2 \rangle = V\, \Delta \nu \left(u(\nu,T) h\nu + \frac{c^3 u^2(\nu,T)}{8\pi\nu^2} \right)$$

The first term $V\, \Delta \nu\, u\,(\nu,T)\, h\nu$ is the total energy E in the volume V and in the frequency interval $\Delta \nu$ multiplied by $h\nu$. If the energy is divided in quanta of energy $h\nu$, there are $E/h\nu = n$ quanta in the volume and the fluctuation of the square of the energy is $\langle \varepsilon^2 \rangle = n(h\nu)^2$. This expression reflects the extreme corpuscular view and is the fluctuation that obtains in the limit of validity of the approximate Wien's law. The second term is the result of the wave interference and obtains in the limit of validity of the approximate Rayleigh-Jeans' law. The exact Planck's law gives the complete formula with both terms.

Appendix 6

Specific Heat of Solids According to Einstein

The law of Dulong and Petit is easily explained by assuming that each mole of a solid element contains A (Avogadro's number) harmonic oscillators. The equipartition of energy gives to each oscillator the kinetic energy $\frac{3}{2}kT$, and an equal amount of energy comes from the potential energy equal, on average, to the kinetic energy, in the case of the harmonic oscillator. Hence the total energy in a mole is

$$E = 3kAT = 3RT$$

and the molar heat is $dE/dT = 3R$.

If the oscillators are quantized their average kinetic energy per degree of freedom is

$$E = \frac{kT \sum\limits_{n=0}^{\infty} nxe^{-nx}}{\sum\limits_{n=0}^{\infty} e^{-nx}} = kT \frac{x}{e^x - 1}$$

where $x = h\nu/kT$. This formula has already appeared in Appendix 2, Equation (10).

The molar heat of a solid is then

$$C = \frac{dE}{dT} = 3R \frac{x^2 e^x}{(e^x - 1)^2}$$

This formula is in fair agreement with experiment but needs important improvement mainly because the frequency ν must be replaced by a whole spectrum of frequencies.

P. Debye evolved a much better approximation that takes into account such a spectrum.

Appendix 7

A and *B* of Einstein

Take a system with two quantum states, r and s, in a black radiation field of the density $u(v,T)$. The number of systems in the state s is n_s and the number of systems in the state r is n_r. Boltzmann had found much earlier that in thermal equilibrium the ratio n_s/n_r is given by

$$\frac{n_s}{n_r} = e^{-\frac{E_s - E_r}{kT}}$$

This is a fundamental law of statistical mechanics called Boltzmann's law.

A system in the state s has the probability per unit time A of spontaneously falling into the state r, and the probability per unit time (proportional to the radiation density) $B\,u(v,T)$ of being pushed into the state r by radiation. Similarly, a system in the state r has the probability per unit time $C\,u(v,T)$ of being pushed by radiation into the state s. The energy $E_s - E_r$ divided by h gives the frequency of the radiation emitted and absorbed by the transition, according to the idea of light quanta. This is also the frequency for which we must to calculate $u(v,T)$. In equilibrium we must have the same number of transitions per unit of time in both directions, and therefore:

$$n_r C\,u(v,T) = n_s[A + B\,u(v,T)]$$

Combining Boltzmanns law and the equilibrium condition, we find:

$$\exp[-(E_s - E_r)/kT](A + B\,u(v,T) = C\,u(v,T)$$

For T tending toward infinity, $u(v,T)$ becomes very large, $A \ll B\,u(v,T)$ and $\exp[-(E_s - E_r)/kT] = 1$, hence we must have $B = C$, and solving for $u(v,T)$ we obtain:

$$u(v,T) = \frac{A/B}{e^{(E_s - E_r)/kT} - 1}$$

From Wien's thermodynamic law $u(\nu,T) = \nu^3 f(\nu/T)$, valid for any system, we must conclude that $E_s - E_r$ is proportional to ν and A/B to ν^3. We have thus a remarkable justification for the relation $E_s - E_r = h\nu$. For $h\nu \ll kT$ the expression for u goes into the approximation $u = AkT/Bh\nu$ and this must coincide with the Rayleigh-Jeans formula $u = 8\pi\nu^2 kT/c^3$. This requirement gives $A/B = 8\pi h\nu^3/c^3$. Hence we have a complete derivation of Planck's law. Conversely if we assume Planck's law, we have a derivation of the value of A/B. This ratio can also be calculated on a specific example, by quantum mechanics.

Appendix 8

J. J. Thomson's Parabola Method for Finding e/m of Ions

Ion beams produced with J. J. Thomson techniques contained ions of vastly different velocities. A method was required that would make it possible to use them for the determination of e/m without losing too much intensity.

Thomson used perpendicular deflections by an electric and a magnetic field. Suppose the ions move in the z direction with velocity v and are subject to an electric field E and a magnetic field B, both in the x direction. The Lorentz force acting on the ion is

$$\mathbf{F} = e\mathbf{E} + \frac{e}{c}\mathbf{B} \times \mathbf{v} \tag{1}$$

and in components, for our case

$$F_x = eE \qquad F_y = \frac{e}{c}Bv \tag{2}$$

the corresponding accelerations $\mathbf{F}/m = \mathbf{a}$ and the deflections in a plane perpendicular to the z axis are:

$$x = \frac{eE}{m}\frac{t^2}{2} = \frac{eE}{m}\cdot\frac{A}{2v^2} \tag{3}$$

$$y = \frac{eBv}{mc}\frac{t^2}{2} = \frac{eB}{mc}\cdot\frac{A'}{2v} \tag{4}$$

Here A and A' are constants depending on the geometry of the apparatus. From Equations (3) and (4) we have

$$y^2 = \frac{e}{m}\frac{B^2A'^2}{2Ec^2A}x = kx$$

Hence the ions land on a parabola irrespective of velocity; the velocity determines which point of the parabola they hit. Each value of e/m gives rise to its own parabola.

Bohr's Hydrogen Atom

Consider an atom formed by center of infinite mass and positive charge e attracting electrostatically an electron. The orbits will be keplerian ellipses with the nucleus (the infinite mass) in a focus. For simplicity we shall consider only circular orbits. Call the mass and the velocity of the electron m and v, its angular velocity ω, and the distance from the nucleus r. Balancing centrifugal force and Coulomb attraction we have

$$\frac{mv^2}{r} = m\omega^2 r = \frac{e^2}{r^2} \tag{1}$$

If the nuclear mass is finite, the same equations obtain, replacing m by $Mm/(M + m)$ (reduced mass), and if the nucleus has a charge Ze, as in hydrogenic ions, replace one of the e by Ze. Henceforth, m is the reduced mass.

The zero of the energy of the system corresponds by convention to the electron at rest at an infinite distance from the nucleus. With this convention the energy of an orbit is

$$E = \frac{mv^2}{2} - \frac{Ze^2}{r} = - \left(\frac{Z^2 e^4 m \omega^2}{8} \right)^{1/3} \tag{2}$$

Equation (2) gives

$$\frac{\left| E^3 \right|}{\omega^2} = \frac{Z^2 e^4 m}{8} = \text{constant} \tag{3}$$

The energy radiated in a quantum jump from an orbit labeled 1, to another orbit labeled 2, appears as radiation of frequency

$$\nu = \frac{E_1 - E_2}{h} \tag{4}$$

The structure of Balmer's formula

$$\nu = R \left(\frac{1}{n_1^2} - \frac{1}{n_2^2} \right) \tag{5}$$

suggests that R/n_1^2 is an expression for E_1/h for specific orbits. We want that for n large, corresponding to large orbits, and for the minimum jump of n (by one unit), Balmer's formula gives a frequency tending to the classical value, that is, to $\omega/2\pi$.

Balmer's formula gives for ν, approximately,

$$\nu = R \left(\frac{1}{n^2} - \frac{1}{(n+1)^2} \right) \simeq \frac{2R}{n^3}$$

ω is given by Equation (3) and combining the two expressions, and replacing $|E|$ by hR/n^2

$$2\pi\nu = \frac{4\pi R}{n^3} = \left(\frac{8E^3}{Z^2 e^4 m} \right)^{1/2} = \left(\frac{8R^3 h^3}{n^6 Z^2 e^4 m} \right)^{1/2} \tag{6}$$

By equating the second and the last term we have:

$$R = 2\pi m Z^2 e^4 / h^3 \tag{7}$$

From the value of R we obtain from Equation (5)

$$E = -\frac{Rh}{n^2} = -\frac{2\pi^2 m Z^2 e^4}{h^2 n^2} \tag{8}$$

The negative values of E belong to the bound states. All positive energy values, which classically belong to hyperbolic orbits, are allowed. They form a continuous spectrum.

The allowed orbits' radii are

$$r_n = \frac{h^2 n^2}{4\pi^2 m Z e^2} \tag{9}$$

The velocity on the orbits is $2\pi e^2 Z/hn$ and the angular momentum as computed from Equations (9) and (6) is

$$l = \frac{nh}{2\pi} \tag{10}$$

The integral value of the angular momentum in units of $h/2\pi$ is characteristic of the allowed orbits and can be used as a quantization condition.

Appendix 10

Quantum Mechanics
in a Nutshell

We give here a summary of the conceptual schema of quantum mechanics. It is obvious that our exposition is incomplete and does not pretend to teach quantum mechanics. The reader who wants an elementary, but more complete presentation, should turn to one of the many available textbooks. F. Mandl, *Quantum Mechanics* (London: Butterworths, 1957), is short and clear.

Consider a system with one degree of freedom, of coordinate q, and call A, B, etc. dynamical variables such as coordinate, momentum, energy. Let us measure immediately one after the other, the three magnitudes A, B, and A, obtaining the values a' and b' for the first two. For the third it may happen that the second measurement of A will necessarily always give the result a'. We then call A and B compatible observables; otherwise, they are called incompatible observables. When we have made sufficient measurements on a system so that any further measurement shall be either incompatible with the preceding ones or give a result calculable with certitude from the preceding measurements, we have accomplished a complete observation of the system. A complete observation defines the "state" of a system.

To each state of a system there is associated a $\psi(q,t)$ that changes in time according to Schrödinger's equation

$$H_{op}(p,q)\,\psi(q,t) = \frac{h}{2\pi i}\frac{\partial \psi}{\partial t} \qquad (1)$$

where $H_{op}(p,q)$ is the Hamilton operator obtained from the expression of the energy as a function of q and its conjugate momentum p, by replacing p with the operator $(h/2\pi i)(\partial/\partial q)$. For instance, in the case of an harmonic oscillator in one dimension the potential energy is $\frac{1}{2}kx^2$ and the kinetic energy is $\frac{1}{2}p^2/m$. Inserting in Equation (1) we have

$$H(p,x)\,\psi(x,t) = \left(-\frac{h^2}{8\pi^2 m}\frac{d^2}{dx^2} + \frac{1}{2}kx^2\right)\psi(x,t) = \frac{h}{2\pi i}\frac{\partial \psi(x,t)}{\partial t} \qquad (2)$$

It is clear that given the initial $\psi(q,0)$ one can find the $\psi(x,t)$ at any time by integrating Equation (1).

We also have the following postulates:

1. The result of the measurement of the physical observable $G(q,p)$ is one of the eigenvalues of the equation

$$G_{op}(p,q)\varphi_n(q) = g'_n\varphi(q) \tag{3}$$

As a particular case, one has for the energy

$$H(p,q)\psi_n(q) = E_n\psi(q) \tag{4}$$

which is Schrödinger's equation. If, as an example, we have measured the energy of an oscillator and found the value E_3 corresponding to the eigenvalue of $\psi_3(x)$, then we know that the $\psi(x)$ associated with the state of our oscillator is $\psi_3(x)$, specifically E_3 is $(3 + \frac{1}{2})h\nu$ with $\nu = (1/2\pi)\sqrt{k/m}$.

2. The state of a system is defined by its $\psi(q,0)$ as stipulated above. The $\psi(q,0)$ may be developed in a series of eigenfunctions of $G(q,p)$ by the Fourier process obtaining

$$\psi(q) = \Sigma a_n\varphi_n(q) \tag{5}$$

3. If one measures G in the state ψ the probability of finding the eigenvalue g'_n is $|a_n|^2$.

The same scheme may be expressed in a different form; a system is described by a complex vector of unit length in an abstract space called *Hilbert space*. This space may have an infinite number of dimensions. The state vectors move in Hilbert space according to a deterministic law given by a generalization of Schrödinger's equation (in essence Equation (1)). When we want to know the magnitude of certain physical quantities, we must introduce axes in Hilbert space. The axes depend on the type of the observables in which we are interested. The possible values of the observable label the axes, and the modulus square of the projection of the vector state on a given axis is the probability of finding that specific value of the observable. It may happen that this probability is one.

Measurement of a physical magnitude in general alters the state vector and thus introduces a discontinuity in its evolution. In the simplest case, in which the state is defined by one observation only, assume that in a state ψ we measure G with the result g'_n; then after the measurement the state is no longer ψ but φ_n.

The theory of measurement in quantum mechanics has given rise to long discussions that are not yet entirely exhausted.

Bibliography

For dictionaries and works of general interest see *Modern Men of Science* (New York: McGraw-Hill, 1968) and C. C. Gillispie, ed., *Dictionary of Scientific Biography* 14 vols. (New York: Scribners, 1970–1976). Many academies, including the Royal Society of London and the National Academy of Sciences, Washington, D.C., publish series of biographical notes on their deceased members. *Nobel Lectures in Physics* (New York: Elsevier, 1967) and *Les Prix Nobel* (Stockholm: Norstedt & Söner, 1902–   ) contain lectures and biographies of the laureates. B. Maglich, ed., *Adventures in Experimental Physics* (Princeton, N.J.: World Science Education, 1972–   ) contains accounts of important discoveries by the discoverers. H. A. Boorse and L. Motz, eds., *The World of the Atom* (New York: Basic Books, 1966) is an anthology with generous biographical notes. It is an excellent source for many of the most important original works discussed in this book. M. Jammer, *The Conceptual Development of Quantum Mechanics* (New York: McGraw-Hill, 1966) is an excellent general description of the development of quantum theory. E. Whittaker, *A History of the Theories of Aether and Electricity* (New York: Harper & Bros., 1960) is an erudite work very useful for the study of mathematical aspects; it includes periods prior to the nineteenth century.

Chapter 1: Introduction

For an interesting general history see B. Tuchman, *The Proud Tower: A Portrait of the World Before the War, 1890–1914* (New York: Macmillan, 1966), which gives a personal panorama of Europe at the end of the nineteenth century.

For some of the major scientific personalities see R. T. Glazebrook, *James Clerk Maxwell and Modern Physics* (London: Cassel, 1901); S. P.

Thompson, *Life of Lord Kelvin* (London: Macmillan, 1910); C. W. F. Everitt, *James Clerk Maxwell: Physicist and Natural Philosopher* (New York: Scribner, 1975); L. Königsberger, *H. von Helmholtz* (F. A. Welby, trans.) (New York: Dover, 1956); J. Hertz, *Heinrich Hertz* (Leipzig: Akademische Verlagsgesellschaft, 1927); J. W. Gibbs, "*R. J. E. Clausius,*" in *Proceedings American Academy 16,* 458 (1889); O. Glasser, *Dr. W. C. Röntgen* (Berlin: Springer, 1959) (Second edition: Springfield, Illinois, C. C. Thomas 1972); E. Broda, *Ludwig Boltzmann Mensch, Physiker, Philosoph* (Vienna: F. Deuticke, 1955); Lord Rayleigh, *The Life of J. J. Thomson, O. M.* (Cambridge: Cambridge University Press, 1943); R. J. Strutt, Fourth Baron Rayleigh, *Life of J. W. Strutt, Third Baron Rayleigh, O. M. F. R. S.* (Madison, Wisc.: University of Wisconsin Press, 1968); G. L. De Haas-Lorentz, ed., *H. A. Lorentz—Impression of His Life and Work* (Amsterdam: North-Holland, 1957); J. J. Thomson, *Recollections and Reflections* (London: G. Bell and Sons, 1936). G. P. Thomson, *J. J. Thomson and the Cavendish Laboratory in His Day* (London: Nelson, 1967); R. A. Millikan, *The Electron* (Chicago: The University of Chicago Press, 1963); R. Vallery-Radot, *The Life of Pasteur* (Garden City, N.Y.: Garden City Publishing Company, 1937); R. Willstaetter, *From My Life* (New York: W. A. Benjamin, 1965).

D. L. Anderson, *The Discovery of the Electron* (New York: Van Nostrand Reinhold, 1964) is an elementary summary of many subjects discussed in this chapter. The economic and financial basis of physics is analyzed in P. Forman, J. L. Heilbron, and S. Weart, *Physics circa 1900— Personnel Funding and Productivity of the Academic Establishment,* in *Historical Studies in the Physical Sciences,* vol. 5 (Princeton, N.J.: Princeton University Press, 1975).

Chapter 2: H. Becquerel, the Curies, and the Discovery of Radioactivity

For Becquerel see *Comité du Patronage du Cinquantenaire de la Découverte de la Radioactivité* (Paris: Ecole Polytechnique, 1946).

For the works of the Curies: P. Curie, *Oeuvres de Pierre Curie* (Paris: Gauthier-Villars, 1908); I. Joliot-Curie, ed., *Oeuvres de Marie Sklodowska Curie* (Warsaw: Państwowe Wydawnictwo Naukowe, 1954).

For biographies of the Curies see M. Curie, *Pierre Curie* (Paris: Payot, 1924; English translation, New York: Dover, 1963); Eve Curie, *Madame Curie* (New York: Doubleday, 1949); A. Langevin, *Paul Langevin, mon père; l'homme et l'oeuvre* (Paris: Editeurs Français Réunis, 1971).

A general history of the discovery of radioactivity can be found in A. Romer, ed., *The Discovery of Radioactivity and Transformation* (New York: Dover, 1960).

Chapter 3: Rutherford in the New World: The Transmutation of Elements

The basic source for this chapter is E. Rutherford, *The Collected Papers of Lord Rutherford of Nelson, under the Scientific Direction of Sir James Chadwick* (New York: Interscience, 1962–1965): vol. 1, *New Zealand-Cambridge-Montreal;* vol. 2, *Manchester;* vol. 3, *The Cavendish Laboratory.* The collection also contains historical articles. Rutherford's official biography is A. S. Eve, *Rutherford* (Cambridge: Cambridge University Press, 1939). See also L. Badash, ed., *Rutherford and Boltwood, Letters on Radioactivity* (New Haven: Yale University Press, 1969) and O. Hahn, *Vom Radiothor zur Uranspaltung: eine wissenschaftliche Selbstbiographie* (Braunschweig: F. Vieweg, 1962).

Chapter 4: Planck, Unwilling Revolutionary: The Idea of Quantization

For thermodynamics see L. P. Wheeler, *J. W. Gibbs: The History of a Great Mind* (New Haven: Yale University Press, 1951).

A basic source for this chapter is M. Planck, *Physikalische Abhandlungen und Vorträge* 3 vols. (Braunschweig: Vieweg, 1958). Volume 3 contains historical articles.

For the history of the study of the blackbody see H. Kangro, *Early History of Planck's Radiation Law* (New York: Crane Russak, 1976); A. Hermann, *Genesis of Quantum Theory* (Cambridge, Mass.: MIT Press, 1971); T. J. Kuhn, *Black Body Theory and the Quantum Discontinuity 1894–1912* (Oxford: Clarendon Press, 1978).

For the Solvay Councils see M. de Broglie, *Les Premiers Congrès de Physique Solvay* (Paris: A. Michel, 1951).

Chapter 5: Einstein—New Ways of Thinking: Space, Time, Relativity, and Quanta

The works of Einstein have not yet been collected in their entirety. A volume that deals, incompletely, with relativity is A. Einstein, H. A. Lorentz, H. Minkowski, and H. Weyl, *The Principle of Relativity, A Collection of Original Memoirs on the Special and General Theory of Relativity with Notes by A. Sommerfeld* (New York: Dover, 1952).

None of the numerous extant biographies can be considered definitive. Among the best are B. Hoffmann and H. Dukas, *Albert Einstein, Creator and Rebel* (New York: Viking Press, 1972); C. Selig, ed., *Helle Zeit, dunkle Zeit—In memoriam A. Einstein* (Zurich: Europa, 1976); R. W. Clark; *Einstein: The Life and Times* (New York: World Publishing Co., 1971).

Especially interesting is the autobiographical sketch, P. A. Schilpp, ed., *Albert Einstein, Philosopher-Scientist* (La Salle, Ill.: The Open Court, 1949).

For Einstein's role in quantum theory, see A. Pais, "Einstein and the Quantum Theory" in *Reviews of Modern Physics 51,* 861 (1979).

The correspondences with Sommerfeld, Born, and Besso are very informative. See A. Hermann, ed., *Albert Einstein–Arnold Sommerfeld Briefwechsel* (Basel: Schwabe, 1968), A. Einstein, *The Born-Einstein Letters* (New York: Waller & Co., 1971); A. Einstein, *A. Einstein and Michele Besso Correspondence 1903–1955* (Paris: P. Speziali, 1972).

A fine study of Ehrenfest throws light on an important personality who was close to Einstein: M. J. Klein, *Paul Ehrenfest* (Amsterdam: North Holland, 1970).

For Michelson see D. Michelson Livingston, *The Master of Light: A Biography of A. A. Michelson* (New York: Scribners, 1973).

Chapter 6: Sir Ernest and Lord Rutherford of Nelson

Fundamental for the Manchester period is *The Collected Papers of Lord Rutherford,* vol. 2 (New York: Interscience, 1962–1965). See also J. B. Birks, ed., *Rutherford at Manchester* (London: Benjamin, 1962), which also contains reprints of the works of Rutherford, Bohr, Moseley, and others.

For the Cambridge period see *The Collected Papers of Lord Rutherford,* vol. 3. See also Sir M. Oliphant, *Rutherford, Recollections of Cambridge Days* (Amsterdam: Elsevier, 1972) and P. L. Kapitza, "Recollections of Lord Rutherford," *Nature 210,* 780 (1966).

Chapter 7: Bohr and Atomic Models

The first volumes of Bohr's works have been published: L. Rosenfeld, ed., *N. Bohr, Collected Works* (Amsterdam: North Holland, 1972).

A very important document, written from first-hand knowledge, which reveals many aspects of Bohr's complex personality, is S. Rozental, ed., *Niels Bohr, His Life and Works as Seen by His Friends and Colleagues* (Amsterdam: North Holland, 1967).

For x-rays see P. P. Ewald ed., *50 years of X-ray Diffraction* (Utrecht: Oosthoek, 1962); W. L. Bragg, *The Development of X-ray Analysis* (New York: Hafner, 1975); G. M. Carol, *William Henry Bragg, Man and Scientist* (Cambridge: Cambridge University Press, 1978); S. K. Allison, "A. H. Compton, a biographical memoir," in The National Academy of Sciences of the United States, *Biographical Memoirs 38,* 81 (1965).

For Moseley see J. Heilbron, *H. G. J. Moseley: The Life and Letters of an English Physicist, 1887–1915* (Berkeley: University of California Press, 1974).

One of the best among the many articles dedicated to Sommerfeld is "Arnold Johannes Wilhelm Sommerfeld, 1868–1951," in M. Born, *Ausgewählte Abhandlungen* (Göttingen: Vandenhoeck & Ruprecht, 1963).

For the rotating electron see R. Kronig, "The turning point," in M. Fierz and V. F. Weisskopf, eds., *Theoretical Physics in the Twentieth Century, A Memorial Volume to Wolfgang Pauli* (New York: Interscience, 1960); S. A. Goudsmit, "Pauli and nuclear spin," in *Physics Today 14,* (June, 1961); S. A. Goudsmit, "It might as well be spin," and G. E. Uhlenbeck, "Personal reminiscences," in *Physics Today 29,* (June 1976).

F. Hund, *Geschichte der Quantentheorie* (Mannheim: Bibliographisches Institut, 1967) gives a history of the model building written by an active participant.

For Otto Stern see E. Segrè, "Otto Stern. A biographical memoir," in The National Academy of Sciences of the United States, *Biographical Memoirs 43,* 215 (1973).

For Pauli see first of all R. Kronig and V. F. Weisskopf, eds., *Collected Scientific Papers by W. Pauli* (New York: Wiley Interscience, 1964); also, M. Fierz, and V. F. Weisskopf, eds., *Theoretical Physics in the Twentieth Century.*

Chapter 8: A True Quantum Mechanics at Last

In addition to M. Jammer, *The Conceptual Development of Quantum Mechanics* (New York: McGraw-Hill, 1966), see B. L. Van der Waerden, ed., *Sources of Quantum Mechanics* (New York: Dover, 1968).

For de Broglie see *Louis de Broglie, Physicien et Penseur,* (Paris: Albin Michel, 1953).

For Heisenberg see his personal autobiographical account: W. Heisenberg, *Physics and Beyond; Encounters and Conversations* (New York: Harper & Row, 1971) and also D. Irving, *The Virus House* (London: Kimber, 1967).

For Pauli see the Bibliography for Chapter 7.

For Dirac see P. A. M. Dirac, "Recollections of an exciting era," in *History of 20th Century Physics, Proceedings of International School of Physics, Course 57* (New York: Academic Press, 1977).

For Born see M. Born, *Ausgewählte Abhandlungen* (Göttingen: Vandenhoeck & Ruprecht, 1963) and M. Born, *My Life: Recollections of a Nobel Laureate* (New York: Scribner, 1975).

Some of the classic books that summarize the scientific work of the founders of quantum mechanics are W. Heisenberg, *Physical Principles of the Quantum Theory* (New York: Dover, 1930); W. Pauli, "Die allgemeinen Prinzipien der Wellenmechanik," in Geiger and Scheel, eds., *Handbuch der Physik,* vol. 24/1 (Berlin: Springer, 1933); P. A. M. Dirac, *The Principles of Quantum Mechanics* (Oxford: Clarendon Press, 1930); P. Jordan,

Anschauliche Quantentheorie, Eine Einführung in die Moderne Auffassung der Quantenerscheinungen (Berlin: Springer, 1936); E. Schrödinger, *Collected Papers on Wave Mechanics* (London: Blackie and Son, 1928).

The epistemological discussions between Bohr and Einstein are reproduced in N. Bohr, "Discussions with Einstein on epistemological problems in atomic physics" in P. A. Schilpp, ed., *Albert Einstein, Philosopher-Scientist* (La Salle, Ill.: The Open Court, 1949). Traces can be found in the Born-Einstein correspondence cited in the Bibliography for Chapter 5.

S. Rozental, ed., *Niels Bohr, His Life and Works as Seen by His Friends and Colleagues* (Amsterdam: North Holland, 1967) contains descriptions written by Heisenberg, Pauli, and others on the background and "spirit" of Copenhagen. See also: K. Przibram, *Briefe zur Wellenmechanik* (Vienna: Springer, 1963) for the reactions of Schrödinger. T. S. Kuhn, J. L. Heilbron, P. L. Forman, and L. Allen, *Sources for History of Quantum Physics* (Philadelphia: The American Philosophical Society, 1967) is an indispensable reference work for a study in depth. An English version of the Faust of Copenhagen can be found in G. Gamow, *Thirty Years that Shook Physics* (New York: Doubleday, 1966).

Chapter 9: The Wonder Year 1932: Neutron, Positron, Deuterium, and Other Discoveries

O. M. Corbino, "I nuovi compiti della fisica sperimentale," in *Atti della Società Italiana per il Progresso delle Scienze 18,* 1157 (1929) (translation in *Minerva 9,* 528 (1971).

For the discovery of the neutron see: F. and I. Joliot-Curie, *Oeuvres scientifiques complètes* (Paris: Presses universitaries de France, 1961); J. Chadwick, "Possible existence of a neutron," *Nature 129,* 312 (1932); J. Chadwick, "Some personal notes on the search for the neutron." Ithaca, N.Y., *Proceedings of the X International Congress of History of Science* (1962) (Paris: Hermann, 1964).

For the background of the California Institute of Technology, where C. D. Anderson always worked, see R. A. Millikan, *Autobiography* (Englewood Cliffs, N.J.: Prentice-Hall, 1950); C. D. Anderson, "The positive electron," *Physical Review 43,* 491, (1933).

For Majorana see E. Amaldi, "Ettore Majorana, man and scientist" in A. Zichichi, ed., *Strong and Weak Interactions* (New York: Academic Press, 1966).

The proceedings of the seventh Solvay Council are published in *Comptes-rendus du 7e Conseil de Physique Solvay* (Paris: Gauthier-Villars, 1934). They give a very precise description of the state of nuclear physics in October 1933.

For deuterium see H. C. Urey, F. G. Brickwedde, and G. M. Murphy, "A hydrogen isotope of mass 2 and its concentration," *Physical Review 40,* 1 (1932).

Chapter 10: Enrico Fermi and Nuclear Energy

For Fermi first of all see E. Segrè, ed., *The Collected Papers of Enrico Fermi* (Chicago: University of Chicago Press, 1962), which also contains ample historical-biographical information. Laura Fermi, *Atoms in the Family* (Chicago: University of Chicago Press, 1954). E. Segrè, *Enrico Fermi, Physicist* (Chicago: University of Chicago Press, 1970); "Memoral Symposium in honor of E. Fermi at the Washington Meeting of the American Physical Society, April 29, 1955," *Reviews of Modern Physics 27,* 253 (1955).

On the discovery of fission see O. Hahn, *Vom Radiothor zur Uranspaltung: eine wissenschaftliche Selbstbiographie* (Braunschweig: F. Vieweg. 1962); O. Frisch, "The interest is focussing on the atomic nucleus," in S. Rozental, ed., *Niels Bohr, His Life and Works as Seen by His Friends and Colleagues* (Amsterdam: North Holland, 1957); O. Frisch, *What Little I Remember* (Cambridge: Cambridge University Press, 1979).

On Rasetti see T. Nason, "A man for all sciences," in *The Johns Hopkins Magazine 17–4,* 12 (1966).

On Szilard see G. W. Szilard and K. R. Winsor, eds., "Reminiscences of Leo Szilard," in *Perspectives in American History,* vol. 2 (Cambridge, Mass: Harvard University 1968) and L. Szilard, *His Version of the Facts* (Cambridge, Mass.: MIT Press, 1978).

On the artificial elements see E. Segrè, *I nuovi elementi chimici. Chimica nucleare alle alte energie* (Roma: Accademia Nazionale dei Lincei, 1953). An overview of nuclear physics in the 1930s is given in *Nuclear Physics in Retrospect* R. H. Stuewer, ed. (Minneapolis: University of Minnesota Press, 1979).

On the development of atomic energy and the atomic bomb there is abundant literature, not all reliable. Among the best sources are: H. D. Smyth, *Atomic Energy for Military Purposes* (Princeton, N.J.: Princeton University Press, 1945); R. G. Hewlett and O. E. Anderson, Jr., *The New World, 1943–1946* (University Park: The Pennsylvania University Press, 1962); D. Irving, *The Virus House* (London: Kimber, 1967); E. Bagge, K. Diebner, K. Jay, *Von der Uranspaltung bis Calder Hall* (Hamburg: Rowohlt, 1957); S. A. Goudsmit, *Alsos* (New York: Henry Schuman, 1947); M. Gowing, *Britain and Atomic Energy* (London: St. Martin's Press, 1974); N. I. Golovin, *I. V. Kurchatov. A Socialist-Realist Biography of the Soviet Nuclear Scientist* (Bloomington, Ind.: Selbstverlag Press, 1968); A. K. Smith, *A Peril and A Hope* (Chicago: University of Chicago Press, 1965).

On J. Robert Oppenheimer various writers have let their imaginations run wild. One of the best sources, *In the Matter of J. R. Oppenheimer, Transcripts of a Hearing before Personnel Security Board, Washington. April 12, 1954, through May 6, 1954* (Washington, D.C., Government Printing Office, 1954), requires important background information before it can be usefully consulted. Among the books worth consulting: P. Michelmore, *The Swift Years—The R. Oppenheimer Story* (New York: Dodd, Mead, 1969); D. Royal, *The Story of J. Robert Oppenheimer* (New York: St. Martin's Press, 1969). H. York, *The Advisors: Oppenheimer, Teller, and the Superbomb* (San Francisco: W. H. Freeman and Company, 1976) gives an illuminating description of the political battles around the hydrogen bomb.

See also J. S. Dupré and S. A. Lakoff, *Science and the Nation* (Englewood Cliffs, N.J.: Prentice Hall, 1962). S. M. Ulam, "Thermonuclear devices," in R. E. Marshak, ed., *Perspectives in Modern Physics* (New York: Interscience, 1966).

On the important personality of John von Neumann see S. M. Ulam, H. W. Kuhn, A. W. Tucker, and C. E. Shannon, "John von Neumann, 1903–1957," in *Perspectives in American History,* vol. 2 (Cambridge, Mass.: Harvard University, 1968).

Chapter 11: E. O. Lawrence and Particle Accelerators

A very brief history of cryogenics up to 1935 can be found in M. Ruhemann and B. Ruhemann, *Low Temperature Physics* (London: Cambridge University Press, 1937); see also K. Mendelssohn, *Quest for Absolute Zero* (New York: McGraw-Hill, 1966).

A good historical synthesis on accelerators can be found in E. M. McMillan, "Particle accelerators," in E. Segrè, ed., *Experimental Nuclear Physics* vol. 3 (New York: John Wiley and Sons, 1959); and in M. S. Livingston, *Particle Accelerators: A Brief History* (Cambridge, Mass.: Harvard University Press, 1969).

For E. O. Lawrence see L. W. Alvarez, "Ernest Orlando Lawrence. A Biographical Memoir," in The National Academy of Sciences of the United States, *Biographical Memoirs 41, 251,* (1970); H. York, *The Advisors: Oppenheimer, Teller, and the Superbomb* (San Francisco: W. H. Freeman and Company, 1976).

For the history of CERN: M. Conversi, ed., *Evolution of Particle Physics. A Volume Dedicated to Edoardo Amaldi on his Sixtieth Birthday* (London: Academic Press, 1970); E. Amaldi, "First International Collaboration between Western European Countries," in *Proceedings of International School of Physics, Course 57,* (New York: Academic Press, 1977).

Chapter 12: Beyond the Nucleus

For the introduction of Western physics into Japan see K. Koizumi, "The emergence of Japan's first physicists: 1868–1900," in *Historical Studies in the Physical Sciences 6,* 3 (1975).

For Yukawa see H. Yukawa and K. Chihiro, "Birth of the meson theory," in *American Journal of Physics 18,* 154 (1950).

On Occhialini see *Simposio in onore di Giuseppe Occhialini per il XX anniversario del suo ritorno in Italia. Seminario Matematico e Fisico di Milano* (Pavia: Editrice Succ. Fusi, 1969).

The discovery of the nonconservation of parity is described by C. S. Wu and others in *Adventures in Experimental Physics 3,* 93 (1974). For the reaction of Pauli see R. Kronig and V. F. Weisskopf, eds., *Collected Scientific Papers by W. Pauli* (New York: Wiley Interscience, 1964).

C. N. Yang, *Elementary Particles, a Short History of Some Discoveries in Atomic Physics* (Princeton, N.J.: Princeton University Press, 1962) is an elementary introduction to particle physics.

The interesting history of the Rochester Conferences is given in R. E. Marshak, "The Rochester Conferences," *Bulletin of Atomic Scientists* (June 1970).

Articles that present the views of some Japanese physicists can be found in S. Sakata, *Scientific Works* (Tokyo: Publication Committee of Scientific Papers of Prof. S. Sakata, 1977). For a more generally accepted point of view see Y. Ne'eman, "Concrete versus abstract theoretical models," in Y. Elkana, ed., *The Interaction between Science and Philosophy* (Atlantic Highlands, N.J.: Humanities Press, 1974).

For the two kinds of neutrinos see L. Lederman and others, "Discovery of two kinds of neutrinos," in *Adventures in Experimental Physics 1,* 81 (1972).

Chapter 13: New Branches from the Old Stump

The history of some of the most recent discoveries can be found in *Les Prix Nobel* (Stockholm: Norstedt & Söner, 1902–   ).

For part of the history of modern quantum electrodynamics see R. P. Feynman, "The development of the space-time view of quantum electrodynamics," in *Les Prix Nobel en 1965* (Stockholm: Norstedt & Söner, 1966) and F. Dyson, *Disturbing the Universe: A Life in Science* (New York: Harper and Row, 1979).

For masers and lasers see A. L. Schawlow, "From maser to laser," in B. Kursunoglu and A. Perlmutter, eds., *Impact of Basic Research on Technology* (New York: Plenum Press, 1973).

For nuclear physics see C. Weiner, ed., *Exploring the History of Nuclear Physics* (New York: American Institute of Physics, 1972); R. H. Stuewer, ed., *Nuclear Physics in Retrospect* (Minneapolis: University of Minnesota Press, 1979); H. D. Jensen, "The history of the theory of the atomic nucleus," *Science 147,* 419 (1965); M. Goeppert-Mayer. "The shell model," in *Les Prix Nobel en 1963* (Stockholm: Norstedt & Söner, 1964); E. Segrè, "Artificial radioactivity and the completion of the periodic system," *The Scientific Monthly 57,* 12 (1943); G. T. Seaborg and E. Segrè, "The Transuranium elements," *Nature 159,* 863 (1947); G. T. Seaborg ed., *Transuranium Elements (Stroudsburg: Perma, Dowder, Hutchinson & Ross, 1978).*

For the Mössbauer effect see the introduction to H. Frauenfelder, *The Mössbauer Effect* (New York: Benjamin, 1962).

For the macroscopic quantum effects see C. J. Gorter, "Superconductivity until 1940 in Leiden and as seen from there," in *Reviews of Modern Physics 36,* 3 (1964); K. Mendelssohn, *Quest for Absolute Zero* (New York: McGraw Hill, 1966); P. W. Anderson, "How Josephson discovered his effect," *Physics Today* (November 1970); P. W. Anderson, J. M. Rowell, S. Shapiro, and D. Lauderberg, "Observation of Josephson effect and measurement of h/e," *Adventures in Experimental Physics 3,* 45 (1973).

For the transistor see W. Brattain and others, "Discovery of the transistor effect," *Adventures in Experimental Physics 5,* 1 (1976).

A semipopular explanation of recent discoveries in astronomy is found in S. Weinberg, *The First Three Minutes* (New York: Basic Books, 1977). A biographical sketch of H. Bethe is given by J. Bernstein in *The New Yorker,* December 3, 10, and 17, 1979.

A survey of contemporary physics and its areas of research seen by a group of competent observers is A. D. Bromley, *Physics in Perspective* (Washington, D.C.: The National Academy of Sciences, 1972). The views in the book, however, are not universally shared.

For molecular biology see the very personal J. D. Watson, *The Double Helix* (New York: Atheneum, 1968); R. Olby, "Francis Crick, DNA and the central dogma," *Daedalus 939* (1970).

Chapter 14: Conclusions

Any conversation among physicists, especially those of an older generation invariably touches on the themes of this chapter. The very young try to refute them with the facts, thus justifying, even if they are not aware of them, the words of Einstein:

> If you want to learn from the theoretical physicists about the methods which they use, I advise you to follow this principle very strictly: don't listen to their words; pay attention, instead, to their actions. [Einstein in *Mein Weltbild* (Amsterdam: Querido, 1934)]

Another quote from Einstein:

> The scientific theorist is not to be envied. For Nature, or more precisely experiment, is an inexorable and not very friendly judge of his work. It never says "Yes" to a theory. In the most favorable cases it says "Maybe," and in the great majority of cases simply "No.". . . Probably every theory will some day experience its "no"—most theories, soon after conception. [November 11, 1922 in an Album of Kamerlingh Onnes: *Albert Einstein, The Human Side* (Princeton, N.J.: Princeton University Press 1979) p. 18]

Name Index

Subject Index